Commission of the European Communities

Environmental Management in Agriculture

European Perspectives

Proceedings of a Workshop held from 14 to 17 July 1987 at Bristol, United Kingdom, under the aegis of the CEC Land and Water Use and Management Committee

Edited by J.R. Park

Environmental Unit, Agricultural Development and Advisory Service (ADAS), Ministry of Agriculture, Fisheries and Food (UK)

Sponsored by the Directorate-General for Agriculture as part of the Coordination of Agricultural Research Programme

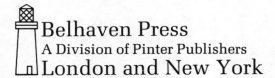

Belhaven Press
A Division of Pinter Publishers
London and New York

© ECSC, EEC, EAEC, Brussels and Luxembourg, 1988

Publication no. EUR 11169 of the Commission of the European
Communities Directorate-General Telecommunications,
Information Industries and Innovation, Luxembourg

British Library Cataloguing in Publication Data

A CIP catalogue record for this book is available from the British
Library

Library of Congress Cataloging in Publication Data

Environmental management in agriculture.

"Prepared for the Commission of the European
Communities."
   Bibliography: p.
   1. Agricultural ecology—Europe—Congresses.
2. Agricultural conservation—Europe—Congresses.
3. Range management— Europe—Congresses. 4. Wetlands
conservation—Europe—Congresses. I. Park, J.R.,
1944-   . II. Commission of the European Communities.
S452.E58 1988   630'.2'745094   87-35133
ISBN 1-85293-036-5

Typeset by Wordbook of London
Printed by Biddles of Guildford Ltd.

Contents

List of figures and tables

Figures

Tables

Preface

Agriculture has a profound impact on the environment of the European Community since two-thirds of its land area is devoted to agricultural production. In the last four decades Europe's agriculture has undergone a technological revolution which has radically changed farming practices, and the pace of technological progress shows no sign of slowing down. There is growing concern about the effect of such changes on the environment — a concern expressed not only among the urban population but also among those engaged in agriculture.

The Common Agricultural Policy has sustained the development of Community agriculture over more than twenty years, with results that are substantial and positive. The policy is now faced with new challenges; the development of technical and economic factors affecting the agricultural sector continues, the emergence of surpluses has created a burgeoning demand on Community budgets and the need remains to maintain employment and address the wider social objectives of the CAP.

In 1985 the Commission issued a Green Paper on 'Perspectives for the Common Agricultural Policy'. On the basis of the opinions received, the Commission then prepared a set of tentative guidelines which contributed to discussions on the development of relevant policies. These guidelines recognised that agriculture, in addition to its economic function, has an increasingly important role in contributing to 'the safeguard of the environment and the countryside'. To this end it was deemed necessary to develop a better understanding of the interaction between agriculture and the environment by intensifying research efforts, particularly with the aid of developing agricultural management practices that reduce the environmental impact of farming systems and land use systems that conserve the countryside and enrich its flora and fauna.

Within the framework of the Community programme for Co-ordination of Agricultural Research, the Programme Committee for Land and Water Use and Management initiated discussions on such research needs in 1986. A first meeting on 'Impact of Agriculture on the Environment', hosted by Denmark, identified a number of topics for further study. As a result the Commission invited the United Kingdom to organise a workshop on 'Agricultural Management and Environmental Objectives'. The aim of the Workshop was to review information available on agricultural management systems for selected habitats capable of meeting production, nature conservation, landscape and other environmental objectives.

The Workshop was held from 14 to 17 July 1987 at Bristol. Siting it in the South West of England based the Workshop in an area of outstanding conservation interest. There are two National Parks, a multitude of sites of special conservation or scientific interest and two recently designated Environmentally Sensitive Areas. Hence, there is extensive local experience of identification of land with environmental interest and agricultural management to achieve environmental objectives. A very useful account of this experience was presented by Mr Halliday (see Introduction, following) and participants saw practical management techniques at first hand.

The Workshop reviewed relevant research in Member States, set out to encourage collaboration on research programmes between Member States, identify gaps in research knowledge and recommend the means to fill these gaps.

The Workshop concentrated on the following topics:

(i)   *management of grassland*: areas used for agricultural production from grazing, and hay or silage making, excluding extensive rangelands and rough grazings on heath and moorland;

(ii)   *management of wetland*: areas, whether upland, lowland, coastal or inland, where there is a permanently high water-table, or which are liable to seasonal water-table fluctuations with or without flooding;

(iii)   *management of field margins*: the areas at the edge of cultivated land, including the adjacent part of the crop as well as the field boundary itself;

(iv)   *monitoring landscape and wildlife habitats*: techniques and systems for monitoring changes in landscape and wildlife habitats related to agricultural practices in terms of the type, location and rate of change.

Member States were requested to provide national reports of the situation in their countries in relation to these topics; seven reports were submitted. Participants were also requested to present research papers.

Review lectures which introduced the technical issues, were presented to the Workshop. Two days were dedicated to intensive discussion of the papers and the intervening day was devoted to a field visit to the Somerset Levels and Moors. The programme is contained in Appendix I.

This publication provides a report of the proceedings and comprises the main points of the introductory talk, the national reports, the review papers, the technical papers and introductions to the parts by the Rapporteurs of each session. The Workshop also made a number of recommendations to the Commission.

Thanks are due to all who took part, to those who assisted in the organisation, particularly the Rapporteurs, and to the Commission for providing financial support for the Workshop. The willingness of participants to contribute actively to the formulation of recommendations, in many cases being required to communicate in a foreign language, is appreciated. The editing of the proceedings has made the minimum changes necessary to improve understanding of the text and has set out to retain the essential flavour of the original.

J.R. Park
Environmental Unit,
Agricultural Development and Advisory Service (ADAS),
Ministry of Agriculture, Fisheries and Food,
July 1987

Introduction

B. Halliday, OBE, Lynton, Devon

My task is to give a brief account of the development of our national approach to agricultural and environmental issues during the last twenty years from the perspective of a farmer in the South West.

Generally speaking, the region is an area of livestock farming with relatively few large conurbations and hardly any heavy industry. The dominant land use in the region is agriculture, but tourism is also important; there are several large areas of relatively unproductive land of poor soil, which have perhaps been less affected by the advance of agricultural technology. Two large stretches of heather and *molinia* moorland, Exmoor and Dartmoor, have been designated as National Parks and much of the land between them is incapable of high levels of production. In addition, a low flat area between Bristol and Taunton, known as the Somerset Levels, is liable to flooding and can only be farmed by means of extensive drainage operations. It has always been a refuge for migrating birds and a reservoir of aquatic and amphibious species of flora and fauna. At the very tip of the south-western peninsular, in Cornwall, there are the relics of an ancient neolithic culture — bronze age field patterns and the relics of an old Celtic language. Both of these areas have recently been designated as Environmentally Sensitive Areas.

National government policy determines which land use should have priority and until comparatively recently, the priority in this country as a whole has been to maximise agricultural production. Now we are beginning to give more priority to the protection of natural habitats and traditional landscape features.

In 1949, growing demand on the land for public recreation was recognised by the designation of Exmoor and Dartmoor as National Parks in which the landscape was to be preserved from change in the

This is an edited version of the introductory talk given by Mr. B. Halliday.

interests of public recreation. These National Parks were to be administered by local authorities under the supervision of the Countryside
Commission. Farmers and landowners did not greet this development
with any enthusiasm. They pointed out that the land was neither
national nor a park: it was their workplace. Thus the seeds were
sown of future political conflict, which began to intensify in the early
1970s. In 1973 the Countryside Commission produced the concept of a
management agreement by which owners of land 'would undertake to
manage their land in a specified manner in order to satisfy a particular
public need.' At about this time I inherited an agricultural estate of
about 600 hectares on the north coast of Exmoor National Park. It
soon became obvious that my small area contained in miniature
many of the unresolved problems of land use which the authorities
were facing on a much larger scale. It seemed a golden opportunity
to try to discover whether it was possible in practice to manage the
land in a balanced way so that it might equally satisfy the interests
of production, public recreation and wildlife conservation. Various
compromises were suggested and a flexible policy of land management
was drawn up. This was to be implemented by a detailed list of agreed
operations; the agreement was signed early in 1979. Certain features of
the agreement were important precedents: compensation in the form
of an annual income supplement was, for example, a new departure.
The need to manage land in a balanced way to reconcile conflicting
interests had already been much discussed in theory but not tried out
systematically in practice. We were also able to demonstrate that, by
co-operative effort, costs could actually be reduced.

Meanwhile, the whole subject of conservation had a higher political
profile. The government EC passed the Wildlife and Countryside Act in
1981; this rejected compulsory powers and required authorities to
achieve their ends by management agreements on a voluntary basis.
A number of threatened species were specifically protected in the new
Act. Shortly afterwards the Nature Conservancy Council proposed to
designate a large area of West Sedgemoor, on the Somerset Levels,
as a Site of Special Scientific Interest (SSSI). National guidelines on
how to finance agreements were published in 1984, and the Nature
Conservancy Council worked extremely hard to arrest agricultural
development on the most important habitat areas. By 1986, ten
Sites of Special Scientific Interest had been notified and a hundred
agreements had been negotiated with individual farmers. In addition,
they established National Nature Reserves over 280 hectares and
acquired or leased a further 400 hectares.

In March 1985, EC Member States approved Regulation 797/85
in which Article 19 permitted Member States 'to take measures to

introduce special national schemes in environmentally sensitive areas (that is, areas of high ecological and landscape value) with.the objective of maintaining farming practices which are compatible with the requirements of protecting the countryside and ensuring an adequate income for farmers.' The British government grasped the opportunity. Section 18 of the Agriculture Act 1986 permits the Minister of Agriculture, after consulting with the Nature Conservancy Council and the Countryside Commission, to designate Environmentally Sensitive Areas (ESA). This Act specifies that each ESA must be a discreet and coherent area capable of being administered as a whole.

Initially, nine areas have been selected; five of these are in England. A prescription for managing the land in each area is laid down and farmers within the ESA are invited to follow this in return for a management payment. On the Somerset Levels the prescription includes limits on cultivation, stocking rates, drainage, on the use of pesticides, herbicides and fertilisers and on the technique and timing of hay and silage cuts. In addition, there is an obligation on the farmer to maintain such features as permanent grassland, pollarded willows, ditches and ditch water levels and certain historic features.

Certain features of this scheme indicate that we have learnt from the mistakes that were made in the development of management agreements. First, the farmer is invited to volunteer to participate and is not obliged to propose a change in order to obtain compensation. Then, under the ESA scheme, the farmer has to manage his whole farm with a view to conserving its whole ecosystem and not just refrain from damaging part of it. But perhaps most important of all, he is given a new constructive role to maintain the traditional landscape.

The response from farmers in these first ESAs has been very encouraging; sufficient, at least, for the Minister to announce that he intends to designate other Environmentally Sensitive Areas. It makes good sense to develop a system of land management aimed at preserving the beauty of the countryside and the survival of its natural flora and fauna which, by its very nature, will call for less intensive food production and thereby contribute to the reduction of food surpluses. But there will inevitably be problems, not least of which will be that of measuring the success of this new method of farming in which the health of the ecosystem is as important as the production of food.

Setting the Scene

1 Efficiency of agricultural industry in relation to the environment

P.B. Tinker,   Natural Environment Research Council, Swindon

There are very few industries of any size or complexity which do not damage their environment in some way or other. Even an office block may be hideous, and industrial buildings will affect their neighbourhood by noise, smell, traffic, appearance or pollution. Britain has a very bad record here, and as a Lancashire man by origin I can still remember the appalling smoke and dirt in the industrial parts of Britain twenty or thirty years ago. Concern on this front is relatively recent, thus Trevelyan's *English Social History* (1973 edition) does not include either pollution or environment in its index.

A major change has taken place in our attitude to environmental factors now, in that they are expected to be foreseen and dealt with before they become significantly damaging. Earlier the approach was to wait until some environmental damage had reached disaster proportions, after which there was a public outcry, and then action could be taken. This attitude is no longer adequate. Recently Jeremy Rifkin, the American activist, said that he wished that genetic engineering should be 'the first technology in modern history to be scrutinised before it came on line'. This may well be true of the release of genetically engineered micro-organisms, which will certainly become a major environmental issue in agriculture in the future, but a similar attitude is now spreading to all other activities also.

The best example of this is in the Environmental Impact Assessment (EIA) approach. The United States pioneered these, but they are now very widely used by major organisations before large projects are implemented. The EIAs are not yet compulsory, but it is expected that

The author wishes to acknowledge the advice given by Dr M. BGell on this chapter and is grateful to Dr Vaidyanathan for permission to use unpublished material.

they will be required for many forms of development in the very near future within the EC.

The situation of agriculture

Until relatively recently, agriculture was almost free of environmental pressures and issues. It was assumed that farming was the function of the countryside, and hence could not harm it. The general approval of agriculture in the public mind also led to the fact that planning constraints on agricultural operations or agricultural buildings are not very restrictive, and that there has been a very strong tendency to take a voluntary approach where some control was necessary. In this sense, agriculture may psychologically be behind other industries, which have accepted supervision relating to buildings, nuisances and pollution. In fact, agriculture has at least one good example of the acceptance of environmental control. In the late 1950s, serious declines in the populations of several birds of prey were seen. Organochlorine insecticides were implicated and their use was restricted gradually, until in 1983 they were banned completely. A decline in the observed content of the organochlorine degradation products in bird livers then followed, and the populations have recovered (Newton and Haas, 1984). Nevertheless it should be recognised that this largely voluntary process took a number of years to carry through, whereas if the same thing happened now, one suspects that there would be an immediate and obligatory ban placed upon such materials.

However, in some respects the earlier views, that agriculture was in some way different, have lingered. For example, the Environment Committee of the House of Commons (1987) noted their surprise at the fact that a farmer may defend himself against a charge of stream pollution by showing that he acted in accordance with 'good agricultural practice'. They concluded 'As a chemical company in this country is expected — indeed required — to safely dispose of its toxic wastes, we see no reason why agriculture should not be expected to have safe slurry and silage stores and disposal routes'.

This changing attitude towards agriculture, with a much less permissive approach, goes together with a generally lowered prestige for the industry. It has received serious criticism for its dependence on subsidies, for the costs of the CAP, and for what many people perceive as being the damage it does to the countryside. A whole list of recent publications could be quoted to support this thesis (e.g. Shoard, 1980). This in part arises from the enormous increase in the number of people who are enrolling in organisations which have strong countryside links, and which therefore may be assumed to be concerned about its condition; these people are in a sense keeping agriculture under observation all the time (Table 1.1). These changes and pressures have combined to a point where a heady brew of suspicion and antagonism is only too possible, and where any major new environmental failure on the part of the agricultural industry will certainly raise the most intense outcry. As I have argued above, the agricultural industry is now in effect being expected to consider future changes very carefully, and to prevent environmental damage occuring, rather than simply responding to it afterwards.

I make this point so strongly because of the almost unpredictable situation in which the industry stands, driven both by the continuous innovations of technology and the unstable economics of the Common Agricultural Policy. There are obviously going to be major changes over the next decade, some of which are very difficult to foresee. For most people habit is overriding, and the population at large has come to accept certain situations in the countryside as being normal. In fact, no part of our countryside is natural, and many treasured features are of relatively recent development. Nevertheless, these are now regarded as the desirable and the normal, so that whatever the changes are, they are almost certain to cause opposition from one quarter or another. It is therefore very appropriate that this meeting includes 'environmental

Table 1.1 Membership of countryside recreation organisations in the United Kingdom

	Membership in thousands		
	1950	1977	1984
Country Nature Conservation Trusts	0.8	115.3	115.0
Royal Society for the Protection of Birds	6.8	244.8	380.0
National Trust	23.4	613.1	1,100.0
Rambler's Association	8.8	29.5	44.0
National Federation of Anglers		446.1	
Royal Yachting Association	1.4	52.1	
British Field Sports Society	27.3	55.0	
Wildflower's Association		34.4	
British Horse Society	4.0	25.5	
Pony Club	20.0	49.5	

Source: *Digest of Countryside Recreation Statistics*, CCP 86, Countryside Commission, Cheltenham, UK, 1978

objectives' in its title, because it is quite clear that very careful planning of our objectives, for a considerable period ahead, is necessary.

Existing controls

The battery of controls over environmental factors in the use of land is increasing very rapidly (Bell, 1987). The environmentally sensitive areas (ESAs) are the most recent, and form a particularly interesting tool in the control of our countryside environment. However, there are now many different designations of areas which are special in one way or another, and one wonders if the proliferation is not becoming confusing. The hopeful part of this proliferation is that it shows that Government accepts a responsibility for maintaining the environmental value of the countryside. Further, there is a strong implication here that some of the large sums of money which currently go to agricultural support (much of which goes to keep material in store rather than to the farmers) could well be re-directed in ways which would have clear environmental value. There is a large amount of money flowing into agriculture at the present time: the task seems to be to re-define objectives so that this flow can be re-directed and thus produce desired, rather than unwanted, results. However, conservation and the environment should not become simply a vehicle for agricultural income-support (Potter, 1986), and considerable regional flexibility will be needed (CAS, 1986).

The relationship between agricultural production and environmental objectives

A number of commentators have considered the likely development of agriculture in the next few years. Broadly there are two extreme scenarios (see Green, 1986; Barber and Ragg, 1987). They are:

1. A sharp division between the agriculturally productive parts of the United Kingdom and the rest. There will be an area of efficient intensive farming, where output, efficiency and profits would be maximised, though presumably avoiding the grossest forms of environmental damage. The rest of the agricultural land would then be available for alternative uses: low intensity farming, leisure, amenity, forestry and all the other suggestions that have been put forward recently. The assumption would be that the remaining agriculture in these areas would be of low productivity and basically unprofitable, but would be supported by Government, for mainly environmental objectives.
2. In the second scenario, it is envisaged that virtually all of the present agricultural area, except perhaps very marginal land, will remain

basically agricultural. However, environmental objectives will be used to control the way in which agriculture is carried out, and this will lessen total production. We must assume that this is done by the re-orientation of the current agricultural subsidies. This seems to be behind the frequent calls for a 'less intensive agriculture' (Green, 1986) or 'low input agriculture'.

It is not difficult to understand the first scenario, though this differentiation between different parts of the country could cause serious strains. It would certainly cause enormous disruption in land values, social behaviour and population.

The second scenario seems to me to be much less clear, and I am not aware of any fully quantitative discussion of it. The underlying assumption appears to be that the adoption of environmental aims, and their grafting onto current agriculture, will thereby reduce yields per hectare in such a way that the total production will fall towards a value which is roughly in balance with demand. In this second scenario of a fundamentally unchanged agriculture, one may consider that the agricultural industry is being asked to attain a whole list of desirable objectives simultaneously. I have tried my hand at producing such a list (Table 1.2), but clearly one could add further points. An industry which can simultaneously meet all these often conflicting demands will indeed be doing a good job. The essence of the question is whether meeting environmental objectives (nos. 6, 7, 8 and 9) will reduce output (no. 1), without doing too much damage to the rest (nos. 2, 3 and 4). Aim no. 5 will not be discussed here.

Environment–agricultural interactions

The decrease in national yield, and the attainment of environmental gain are quite separate objectives, which we could aim for independently. The interest in this approach lies in the possibility that a single action

Table 1.2   'Instructions to Agriculture'

1.	Reduce overall output to meet demand
2.	Reduce cost per unit output, and therefore prices
3.	Maintain quality, availability and variety of food
4.	Maintain the profitability of agriculture as far as possible, so as to minimise the need for subsidies
5.	Maintain employment and the social structure in the countryside
6.	Reduce pollution
7.	Improve countryside appearance
8.	Conserve and increase wildlife and habitat
9.	Improve access to countryside

may bring simultaneous benefit in both respects. It is sometimes assumed that this is so, but it is by no means obviously true of many of the environmental issues. If there is a negative correlation between environmental gain and production, how large is it? If there is a simultaneous loss of production, and environmental gain, how does this affect the farmer's profit, and how much does it cost overall? These are often difficult questions, but any form of the 'second scenario' must address them in detail. The situation may be found far less advantageous than we now assume.

Another point to keep in mind is the reversibility or otherwise of any environmental damage. The degree of public outcry occasioned by an environmental issue is by no means always in proportion to its real long-term effect. The most serious soil erosion of uplands, which may be quite irreversible and may alter a catchment ecosystem for centuries, may only raise protests from a few people with local knowledge. On the other hand, smoke from straw burning, which is a purely temporary nuisance, can cause an enormous outcry. We should take as one of our guidelines the degree to which we are doing permanent, as opposed to purely temporary, harm to the environment. I will discuss some specific issues in relation to these points in the following sections.

1. Field appearance

The lush, uniform green surface of a modern cereal crop, traversed by tidy and parallel tramlines, may bring great satisfaction to the farmer, but many non-farmers do not like it. They prefer variety and colour, with poppies and other weeds flowering freely. Several of the chapters in this book will address this question of diversity of plants within an agricultural crop. It is difficult to decide on such a subjective matter as appearance, and all that researchers can do is to draw up costed at options, in which the degree and style of diversity has its cost in terms of loss of yield and loss of profit carefully identified. Some modern systems, such as encouraging diversity and flowering plants around the field margin, may carry advantages, but a square of solid green set in a neat frame of poppies may look rather artificial. It may be difficult to find a good solution to this problem.

2. Wildlife conservation

This ranges from a reduction in wildlife density, to the final extinction in the United Kingdom of certain species. Where the species in question acts as a pest this is of course a difficult problem, as the economic costs may be very considerable. However, in a general sense, diversity of species is felt to be desirable both by the professional ecologist and by the rambler, and this should follow from more careful attention to the creation of habitat within the countryside. In some cases there may be the need for artificial re-introductions, as with the Adonis Blue butterfly. Additional creation or preservation of habitat will nearly always involve some costs. Bury (1985) has assessed these, and concluded that the capital and maintenance costs were quite small; on ten farms they amounted to £0.55 to £2.93 per ha per year, over a ten-year period. However, opportunity costs in terms of lost production were larger — 2 per cent to 15 per cent of gross total income. Clearly there is opportunity for differences of opinion, in that habitats may be developed in small, inconvenient or sloping pockets of land where the farmer had little incentive to grow crops. However, this issue is important, in that it is a very clear case of environmental benefit being obtained at the same time as a loss of total production, with a quantifiable cost to the farmer.

3. Landscape appearance

This again is a subjective matter, which is affected by a wide variety of other issues. Broadly, this will imply smaller and less intrusive buildings, smaller and less regular fields, more and longer hedgerows and verges, more careful maintenance of hedgerow and other trees. These points may well cause inconvenience and loss of profit, but it is not immediately obvious that they carry a great penalty in yield. The loss of hedgerows is a crucial component of it. In six years, 28,000 km of hedgerow has been lost (Barr et al., 1986). The maintenance of a reasonable density of hedgerows must surely be a prominent environmental objective.

A major issue must be agricultural buildings. Many buildings such as silos or modern barns are grossly obtrusive. Buildings must clearly be functional, but it should be possible to disguise them to some degree, and the cost of better control of buildings in the countryside should not be excessive in terms of yield or profit.

4. Change in soil fertility

Many wild species have poor competitive ability at high soil fertility. Much of our grassland in former times was of very low chemical fertility, because there was a constant drain of this fertility back into the cultivated areas through stocking and manuring management. Once phosphate has been added to a soil in any quantity, its phosphate status will have been raised for many decades. Its correction could indeed be quite expensive, if it implied that farmers should run their land at low phosphate levels, which in modern farming are quite unacceptable. The only compromise appears to be to ensure that fertilizer does not find its way onto uncropped land, so that a farm does retain verges, corners and odd pieces which have a low fertility.

5. Erosion

This is the most obvious and the most permanent damage to the environment, short of the total extinction of an organism. The extent in Britain is small compared with many other countries, but there is a feeling that it is increasing, and some parts of the uplands are exposed to strongly erosive conditions (Morgan, 1985; CEC, 1987). Its prevention should be part of normal good husbandry, and this would certainly imply no loss of yield, but rather the reverse. The Soil Survey of England and Wales has assessed 40 per cent of the land in England and Wales as being potentially liable to erosion under unsuitable management.

6. Heavy metal pollution

Heavy metals are applied to all parts of the country by atmospheric deposition, but specifically to agricultural land by the use of sewage sludge (MAFF, 1980). The immediate dangers are well known, and there are well-defined guidelines for the amounts of heavy metal that may be added. For all practical purposes this is a form of pollution that is permanent, and the setting of acceptable guidelines for heavy metal tolerance therefore needs the greatest care. The effects on agricultural yields of preventing this form of pollution are very small, and may even be positive.

7. Pesticides

These have extremely variable residence times in the soil. For example, DDT has long residence times, and despite the reduction in use, the content in bird livers is still considerable. On the other hand, many organophosphorus insecticides have very short life-times before they are broken down. Their use disturbs the food supply for other organisms. We should encourage in any way possible the more controlled and limited use of insecticides, but it seems unlikely that environmental benefit will demand a massive reduction in their present use. Fungicides are now firmly established as essential for the highest yields in cereals, with a response of roughly 10 per cent in many situations. However, fungi do not appear to have many supporters, so I am not aware of any pressures for banning fungicides. The most important chemicals in this context are probably the herbicides. Their use goes directly against the aim of diversity, but modern intensive farming would scarcely be possible without selective and contact herbicides. There are serious practical problems to be worked out here in terms of acceptable techniques which maintain yield, whilst avoiding the suppression of all environmentally desirable species. Broadly speaking, pesticides do not cause irreversible damage, but any restriction on their utilisation carries serious implications for both yield and profit.

8. Nitrate pollution

Nitrogen is a classic case of the agriculture–environment interaction. In one way nitrate is a most fugitive pollutant, in that it disappears from the soil very rapidly. In another sense it is extremely permanent, in that a nitrate load in an aquifer appears to persist for decades. Nitrate usually moves down the soil and underlying strata at a rate of 0.5–1.0 m per year under British conditions so that it is often a long time before it reaches the aquifer. I will not discuss here whether nitrate does in fact carry any significant health dangers; the EC has laid down a directive on acceptable levels of nitrate, to which we will have to adhere to a greater or lesser degree.

It is now extremely probable that the increasing load of nitrate in surface and groundwaters arises, directly or indirectly, in large measure from increased additions of nitrogen in agriculture (Roberts and Marsh, 1987; Foster et al., 1985). Various schemes for decreasing nitrogen application rates have been put forward. It is very important to establish the optimum level of nitrogen to individual fields more precisely, so that excess amounts are not applied in error. However, the establishment of 'protection zones' around boreholes may be necessary. Within these, the rates of application of nitrogen would be controlled at a low rate, with appropriate compensation for financial loss. The size of the protection zones would need some rather careful research, and would involve an understanding of the way in which groundwater moves in the underlying strata.

Any procedure which limits the utilisation of nitrogen fertilizer would have drastic effects on yield. However, the curved nature of

Table 1.3 Consequences of reduced N input to cereals: mean data from thirty-six sites in East Anglia

Reduction in yield of below maximum, %	Reduction in N dressing below optimum, %	Net loss from reduction, £ per ha
5	40	25.2
10	57	61.8

Source: Private communication from Dr L. Vaidyanathan

the response function means that there is no direct proportionality between a reduction in nitrogen and loss in yield, and in many cases it may be possible to cut application by 40 per cent or 50 per cent, with a loss in yield of the order of 10 per cent. The average situation in shown in Figure 1.1 (Hubbard, 1985). A useful environmental gain could therefore be obtained with a moderate yield reduction, but there is naturally a financial penalty. This has been calculated in detail for a set of thirty-six field experiments in East Anglia by Vaidyanathan (private communication) (Table 1.3). The loss of profit given in this table is of course considerable, and in many cases will wipe out the profitability of the winter wheat enterprise altogether. It is then a political decision to determine how desirable it is to gain the environmental benefit and reduce the overall yield, and to reorganise subsidies and supports accordingly.

More such information is required to enable clear decisions to be taken in this triple balance between environmental gain, size of yield and farming profitability. This subject appears to require more attention, because the only detailed consideration appears to have been given by Raymond (1985). Similar considerations should apply to grassland, where it is known that grazed grassland when heavily fertilized with nitrogen suffers serious leaching losses of nitrate (Ryden and Garwood, 1984).

9. Organic effluents

There is an increasing number of cases in which organic effluents from farms find their way into water courses, with disastrous results. The extreme toxicity of these effluents arises from their high biological oxygen demand. These are usually point source discharges, and result in local fish kills and damage to other aquatic biota. Such cases, while causing intense annoyance to other users of a river should really be avoided by good husbandry. They are in no sense essential for yields or profits, and are thus in a different class to some of the other environmental damage we are considering. Whereas a serious incident may cause considerable damage to aquatic organisms, the situation is usually reversible, once the immediate point source pollution has been rectified.

10. Straw burning

The most recent results show that there is little if any advantage in straw burning, as opposed to incorporation, in terms of cereal yields, if appropriate techniques are used. There is some penalty in profit, which may be of the order of £25 per hectare, if the straw has to be incorporated

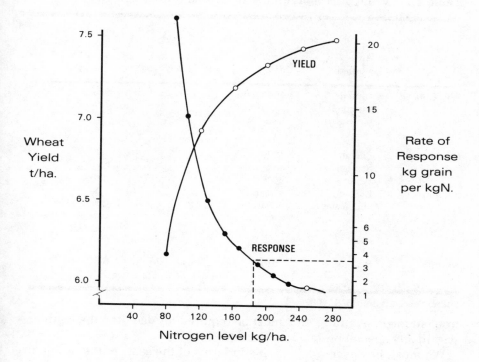

Figure 1.1   The response of winter wheat (after wheat) to different levels of N fertiliser: economic optimum yield 7.26 t/ha at 185 kgN/ha
Source: derived from Hubbard, 1985

instead of burned. Nevertheless, this form of environmental damage is usually more of a passing nuisance than a major problem, despite the enormous irritation that it causes.

Conclusions

It appears from the above discussion that, of the various causes of environmental damage discussed, only the development of wildlife habitat and use of nitrogen and pesticides carries a direct and clear interaction between the environmental consequences, the yield and the profitability of farming. For these we need more exact analyses of the yield and cost effects of adopting specific environment-friendly methods. Quite apart

Table 1.4   Comparison of Studies of farmland requirements

	Area studied	At what date	Area/range (m ha)	Main 'surplus' estimate (m ha)
Wye College	United Kingdom	2000 with sensitivity tests	1–6	3–4
Goulds	England Scotland Wales	1990 and 2000		1.1 2.6
Reading	England and Wales	5 years of various scenarios	0.2- 2.2	1.3 (price pressures) 1.9 (quotas)
North	Britain	2015	5.4 5.72	5.5

Source: derived from Bell, 1987

from strategic planning, this must be required for deciding the regimes within any special protection zones or ESAs.

We must, however, also see this in terms of the scale of the problem of surpluses. Several organisations and authors have made forward predictions of the degree to which technological change will render more agricultural land unwanted in the United Kingdom. A summary of these estimates is in Table 1.4 (Bell 1977). These range up to an estimate of land surplus of 5.5 m hectares in twenty-eight years' time, though no doubt a large error attaches to them. Nevertheless, they give some indication of the scale of the problem. If we accept 'scenario 2' and assume that in practice the whole present area of agricultural land will remain in production, it implies that environmental constraints, field fragmentation, new forest planting, hobby farming, and other such effects, will bring about a quite radical reduction of mean production rate, compared with what technology makes possible at that time in the future. My own feeling is that such a self-imposed degree of inefficiency in utilisation of farmland is unlikely, and that if this is the real scale of the surplus problem, there will gradually be a concentration of intensive agriculture in a limited part of the country, as in 'scenario 1'. Quotas on production may be introduced, so that 'scenario 2' is in effect imposed for particular crops, but if these quotas are saleable between farms, this may mean the gradual drift of intensive agriculture into the most favoured areas. We should then end up with the country roughly partitioned into areas of high agriculture, and other uses.

Amidst the problems and stresses of the present time, there should be opportunities. Technological advances do offer the possibility of major environmental gain and it is the task of those involved in this subject to determine the various ways in which they can be taken.

References

Barber, Sir D., and Ragg, S. 1987. Changes in land use. In *Farming into the 21st Century*, Norsk Hydro Fertilizers, pp. 7–33.

Barr, C., Benefield, C., Bunce, R., Riddsdale, H., Whittaker, M. 1986. *Landscape Changes in Britain*, Institute of Terrestrial Ecology, Monks Wood, Huntingdon.

Bell, M. 1987. *The Future Use of Agricultural Land in the United Kingdom*. Agricultural Economics Society Conference, Reading (in press).

Bury, P). 1985. Agriculture and Countryside Conservation, Farm Business Unit occasional Paper No. 10, Department of Agricultural Economics, Wye College, University of London.

CAS. 1986. *Countryside Implications for England and Wales of Possible Changes in the Common Agricultural Policy*. Centre for Agricultural Strategy, Reading.

CEC. 1987. *The State of the Environment in the European Community 1986*. Commission of the European Communities. Luxembourg.

Environment Committee of the House of Commons, 1987. *Third Report, 1986–87*, pp. 41.

Foster, S.S.D., Geake, A.K., Lawrence, A.R. and Parker, J.M. 1985. Diffuse groundwater pollution: lessons of the British experience. Memoirs of the 18th Congress of Hydrogeologists, Cambridge, pp. 168–77.

Green, B. 1986. Agriculture and the environment. *Land Use Policy*, 3, 193–204.

Hubbard, K. 1985. Examination of the optimum input in cash crop production. In Raymond, ed., *Tuning Inputs to Maintain Profits*, pp. 8–15.

MAFF. 1980. *Inorganic Pollution and Agriculture* (Reference Book 326) HMSO, London.

Morgan, R.P.C. 1985. Assessment of soil erosion risk in England and Wales. *Soil Use and Management*, 1, 127–31.

Newton, I. and Haas, M.B. 1984. The return of the sparrowhawk. *British Birds*, 77, 47–70.

Potter, C. 1986. *The Countryside Tomorrow: A Strategy for Nature*. The Royal Society for Nature Conservation, Lincoln.

Raymond, W.F. ed. (1985) *Tuning Inputs to Maintain Profits*. The Society for the Responsible Use of Resources in Agriculture on the Land, Edenbridge, Kent, 1985.

Roberts, G. and Marsh, T. 1987. The effect of agricultural practices on the nitrate concentrations in the surface water domestic supply sources of western Europe. *Water for the Future: Hydrology in Perspective*. IAHS Publ. no. 164, 365–80.

Ryden, J. and Garwood, E. 1984. Revaluating the nitrogen balance of grassland. In Hardcastle J.E.Y. ed, *Grassland Research Today*, London, Agricultural and Food Research Council, pp. 10–11.

Shoard, M. 1980. *The Theft of the Countryside*. Temple Smith, London.

Trevelyan, G.M. 1973. *English Social History*. Longmans.

2 Monitoring of landscape and wildlife habitats

A.J. Hooper Ministry of Agriculture, Fisheries and Food, London

Knowledge and information about the distribution and characteristics of landscape features and wildlife habitats, and any changes that are occurring, are fundamental to the establishment and implementation of strategies for the enhancement and protection of the environment. There is a need for such information not only to assess the impact of policies and strategies, but also to provide a basis for the development of future initiatives. The need applies at international, national and local levels.

Recognition of the need for national datasets and land inventories is not new and reference is made later to the Land Utilisation Survey in England and Wales organised by Stamp in the 1930s. However, during the past ten years interest has been stimulated by growing concern about the rate and scale of some changes in the environment. Greater agricultural efficiency and intensification of farming are invariably cited as the primary causes of change. But non-agricultural factors have also been significant including forestry, urban expansion, road construction, industry and mineral extraction. These pressures and conflicts have provided the stimulus for a number of national programmes in Great Britain designed to provide data on the changes that are taking place, especially in the countryside.

During the same period there have been significant scientific and technological developments in the fields of remote sensing, computing and statistics. The potential of earth observation satellites for resource monitoring was anticipated from the first launch in the early 1970s. The increased spatial and spectral resolution of the latest generation of satellites has enhanced the capability of satellite systems. At the same time the development of computers with vastly improved performances and the development of computer-based geographic information systems have greatly increased our ability to store, manipulate and analyse spatial data. There have also been important and related developments in the statistical field providing new mathematical tools for the analysis of complex biological and spatial data and the modelling of ecological change.

Some of the changes that we may need to monitor are readily apparent on a short timescale. Forest planting or clearance and the ploughing of traditional permanent pastures are obvious examples of gross changes readily observed and recorded. Other changes occur slowly and less obviously but become significant incrementally. They may be geographically dispersed or concentrated in large blocks. Either extreme imposes significant constraints on the design and specification of a monitoring programme, especially when sampling techniques are to be used. The survey and data collection methods that can be used to monitor changes differ according to the type and extent of change or features to be studied. At one end of the scale there is a need for data at the national level to indicate overall trends in, for example, landscape change or habitat loss. At the other extreme there is often a need to monitor change on a local scale, in relation to a particular habitat or type of pressure. Usually the problems encountered in the collection, collation and analysis of data increase with the size and complexity of the area of interest.

This chapter will examine a number of national surveys, although some examples of more localised studies will also be given. The following brief review of selected monitoring programmes is concerned primarily with methodology and technical issues rather than the results obtained.

Monitoring methods and techniques

The selection of a feasible monitoring technique at an acceptable cost will influence the number and type of landscape or habitat classes which can be accommodated. In land resource surveys a distinction should be made between land cover and land use classes. The latter require information about the management or use of the land which may not be available from some survey methods, such as remote sensing.

In monitoring exercises, a useful distinction may be drawn between firstly, the area and occurrence of land types or features and, secondly, changes in the quality, condition or characteristics of these features. There are some obvious similarities between landscape features and habitat types for monitoring purposes. Although it is often said that the overall quality of a landscape amounts to more than the sum of its constituent parts, most attempts to monitor landscape change involve the identification and measurement of changes in the separate elements of the landscape. These elements are often also habitat types. Thus, for example, changes in the length and condition of hedgerows can be significant in both landscape and wildlife terms. Techniques such as aerial photography may therefore be used to record and measure both landscape features and habitat types. These types of measurement are usually much more straightforward than those concerned with changes in the quality, condition or composition of habitats. The latter invariably requires detailed ground-based observation techniques.

National surveys and schemes

Land Utilisation Surveys

Although more strictly concerned with land use than habitats or landscape per se, the Land Utilisation Surveys of Stamp and Coleman are worthy of note here because they involved complete national mapping of land use based on field surveys (i.e. census rather than sampling). Both were carried out by volunteers; the Stamp survey between 1930 and 1938 (Stamp, 1962) and the Coleman survey in the 1960s (Coleman and Maggs, 1968). The logistical problems for such large-scale surveys are immense; not least the collation and measurement of a vast amount of mapped data. It is also necessary to ensure that field surveyors are trained and operate to known standards of accuracy. A recent study carried out by consultants for the Department of the Environment concluded that cost and other practical considerations would preclude a complete inventory of land use in the future. Current land use projects all involve sampling to obtain estimates.

ITE surveys

The system of land classification developed by the Institute of Terrestrial Ecology in the 1970s (Bunce et al., 1980) has been used to estimate changes in selected land cover types and landscape features between 1978 and 1984. The system divides all land into thirty-two classes (see Appendix 2.A). Land is assigned to a class on the basis of an analysis of seventy-six indicator attributes such as climate, geology and topography. In 1978 the classes were 'characterised' by means of field surveys of eight one-kilometre squares drawn at random from each of the thirty-two land classes. Field records were made of vegetation, soils, land use, woodland, hedgerows and other features. The field data could be related to the distribution and extent of the land classes and used to obtain estimates of the areas occupied by different land uses or vegetation types.

Surveys of the one-kilometre grid squares were repeated in 1984 allowing comparison of the landscape at the two dates. Digital mapping techniques were used to measure changes. The classification system was then used predictively to obtain national and regional estimates of changes in, for example, hedgerow length, woodlands and agricultural crops (Barr et al., 1986). Although sampling errors may be large for rare or localised features the errors for widespread features were found to be within acceptable tolerances.

National Countryside Monitoring Scheme (NCMS)

The NCMS was set up by the Nature Conservancy Council to provide data on changes in the distribution and extent of 'structural components' of the countryside since the 1940s, especially those involving important features such as natural and semi-natural habitats. The sample-based scheme is designed to permit local as well as general trends to be identified and aims to obtain comprehensive data for all parts of the country so that comparisons can be made between different areas on a standard basis.

The survey is being carried out on a county-by-county basis. The results for the county of Cumbria were published recently (NCC, 1987). Land cover types are being mapped from standard black and white aerial photographs covering sample grid squares selected on a stratified random basis within three categories of land type, namely: lowland, intermediate and upland. A county is stratified into these three land types by amalgamating appropriate land classes of the ITE classification system. The total area to be sampled (i.e. county) is divided into discrete 5 km by 5 km squares based on the national grid. The sample squares are then selected at random but so as to provide an approximately equal number of squares in each land type. The total area sampled is about 10 per cent, this level of sampling being that considered necessary to meet the specified target accuracy, which is to estimate a change of 10 per cent for a main land cover type with 95 per cent statistical confidence.

The list of features to be recorded (Appendix 2.B) was selected with the feasibility of consistent and accurate photo-interpretation in mind. Photo-keys have been developed to establish criteria for identification which will improve the consistency of feature recognition. As with any survey based on aerial photography there is dependence on the availability of archival photography at prescribed time datum points. Although the availability of black and white cover is generally good, modification of the sampling areas may be unavoidable due to gaps in aerial photographic coverage. This is a potential source of unknown bias which is difficult to avoid in studies of historic land use change.

Land cover data are digitised via a photogrammetric plotting machine enabling area, length and change measurements to be made with known accuracy. The total sample area covered by each land cover type in each

ITE land class is estimated. The results are weighted relative to the abundance of each land class in each broad land type. The weighted estimates are then combined to provide overall estimates for each land type and for the whole county.

The whole programme was originally planned to take ten years although modifications to the scheme are being considered which should shorten this timescale. The scheme has added to knowledge about the problems and practical constraints of large-scale monitoring systems. The accuracy of photo-interpretation has been found to be affected by a number of factors including the quality, scale and time of year of the photography. Consistent discrimination between grassland types is difficult; also between improved grassland and some arable crops. In the survey of Cumbria, it proved difficult to identify felled woodland. The subjectivity involved in drawing vegetation boundaries in the uplands can give rise to spurious indications of change. It was found essential to make field checks in each square to assess the accuracy of photo-interpretation and the survey areas of uncertainty. This is clearly more difficult for old photography. Estimates were found to be less satisfactory for features that occur in large isolated blocks, notably coniferous plantations and standing natural water.

Monitoring Landscape Change (MLC)

The MLC Project was commissioned in 1984 jointly by the Department of the Environment and the Countryside Commission for England and Wales (Hunting, 1986). It was a two year project carried out by Hunting Surveys and Consultants Limited. The principal objectives were:
(a) to obtain reliable information on the current and past distribution and extent of landscape features of policy importance;
(b) to determine the magnitude of any change in distribution and extent of these features between specific points in time, thus defining rates of change and possible bases for the prediction of trends; and
(c) to develop a method by which future changes in the extent of features can be monitored.

The project was designed to make use of aerial photography, satellite remote sensing and field data. Area and linear features were treated separately. No attempt was made to assess the impact of changes on the quality of landscape as such; the project concentrated on changes in selected land cover types and linear features such as fences, hedges and walls.

The final classification of features (Appendix 2.C) comprised seven main categories each being divided into finer sub-divisions. The classification was ultimately a compromise between information requirements and the reliability with which the features could be identified on aerial photographs or satellite images. The classes were intended to be cover types rather than land uses or species of vegetation.

For the estimation of area features stratified random sampling was used to select 707 sites for aerial photographic interpretation. Strata were defined on the basis of major soil groups in each county. Each site was approximately 5 km² in extent. There were at least ten sites per county and two sites per soil stratum. The sampled area amounted to 2.4 per cent of the land area of England and Wales.

The project was designed to compare data at three dates: 1951, 1971 and 1981. In practice aerial photographs approximating to those dates had to be used. Photo-interpretations were compiled on overlays to a common base and boundaries were digitised. Field data were used to assess the accuracy of photo-interpretation. For area features, an overall interpretation accuracy of 87 per cent was calculated. Over 98 per cent accuracy was claimed for major cover types such as woodland, farmed land and developed land. Interpretation was significantly less accurate for certain categories, especially distinguishing between types of woodland, upland heath and grassland and between cultivated land and permanent grass. A summary of area feature results is shown in Appendix 2.D. One cause of mis-classification was related to difficulties in applying precise definitions of the categories. Also, in the case of upland heath some apparent errors may have been due to problems of mapping of vegetation boundaries on the ground. As has already been stated in relation to the NCMS, discrimination of some grassland types proved to be particularly difficult.

A sub-sample of 140 sites was used to estimate change in linear features, selected so that at least two occurred in each county. Each site, centred on an area features site, was 12 km² and the total area sampled covered 1.1 per cent of England and Wales. New photography was obtained to enable changes between 1981 and 1985 to be estimated. The total length of each type of feature was measured from overlays. An accuracy assessment was made by comparing photo-interpretation and field data at fifty randomly selected points in each of twelve sites and overall accuracy was found to be 86 per cent. On the basis of 1,081 observations the mean accuracies for linear categories were: hedges 85 per cent, fences 77 per cent, walls 72 per cent, banks 87 per cent, ditches 95 per cent, woodland fringe 98 per cent and urban boundary 97 per cent. As for area features it was found that some errors were due to problem of definition rather than photo-interpretation.

Thirteen Landsat Thematic Mapper satellite images acquired in 1984 and covering most of England and Wales were also used to estimate the extent of major area features. The analysis was carried out as a census and not by sampling. Computer-aided image analysis techniques were used to classify the images, aided by 'ground truth' data for selected training areas. Automoated image classification routines were then used to extrapolate from training sites to complete scenes. Classification accuracies were found to be greatest for a four-waveband combination (Bands 7, 5, 4, 3). Accuracy was assessed using a measure termed the 'mean accuracy indicator', which takes account of errors of commission and omission. The accuracy for cultivated land was found to be 89 per cent, improved grassland 63 per cent, unimproved grassland 61 per cent, broadleaved woodland 87 per cent and coniferous woodland 93 per cent. There was wide variation in the performance of spectral classifications for individual TM scenes and as would be expected the date of the imagery proved significant. The problems of mis-classification in both space and time were such that satellite data did not prove to provide an independent and reliable source of national information in the present state of the art.

Designated areas

In addition to the national monitoring projects and schemes described above, increased attention is being paid to monitoring of designated areas, that is areas with statutory status such as the National Parks and Environmentally Sensitive Areas. These are areas of scenic and wildlife importance which are both dependent on and 'threatened' by various forms of human activity such as forestry, mining, farming or tourism.

The Countryside Commission for England and Wales is developing and sponsoring a co-ordinated monitoring programme for the National Parks using aerial photography. Sample-based surveys have been rejected in favour of complete surveys. Information about past changes and recent trends will be obtained and contemporary data will be used to establish a baseline against which future change can be measured.

A practical example of a recent detailed assessment of change in such an area is the study of Broadland in Norfolk and Suffolk by Robin Fuller of the ITE (Fuller et al, 1986). Interpretation of black and white aerial photographs at 1 : 10,000 scale was combined with digital cartography to obtain information about the extent and distribution of semi-natural habitats and man-made land uses. Interpretation accuracy was assessed by comparing the interpretation maps with large scale (1 : 2,000) aerial colour slides of 110 sample points spaced on a grid over the area. Based on a sample of 2,253 points within these 110 areas, the overall concordance was 88 ± 3 per cent. The concordance for wetland communities was generally better than 90 per cent, although the score for some rarer classes was lower. Fuller concluded that the accuracy was sufficient to permit a valid detailed examination of the extent and status of semi-natural and man-made habitats in Broadland.

Environmentally Sensitive Areas

The Ministry of Agriculture, Fisheries and Food has recently started a detailed monitoring programme for six newly-designated Environmentally Sensitive Areas in England and Wales. These are areas in which the landscape, wildlife and historic interest is threatened by possible changes in farming practices, usually intensification but in some areas by undergrazing or neglect. Farmers receive annual payments in return for agreeing to farm by less intensive methods, for example, to restrict fertiliser or pesticide use and to avoid damaging landscape or historic features. A combination of 1 : 10,000 scale aerial photography and field checking will be used to establish a complete baseline land cover inventory for the first year of the scheme. Surveys will be repeated to measure any changes in the extent or distribution of key features. Colour infra-red photography is being used for the two 'wetland' areas, the Somerset Levels and Moors and the Broads. Intelligence and data on management practices will be obtained from aerial photography and ground survey. The monitoring of the floristic quality of key habitats, notably grasslands and dykes, is not feasible using aerial photography and will involve sample-based field surveys.

Concluding remarks

The aim in the preceding sections has been to highlight some of the recent operational schemes and projects concerned with or involving monitoring landscape, wildlife habitat or land use in parts of the United Kingdom. Reference could also have been made to many other related activities such as the commissioned research programmes of the Natural Environment Research Council. It should be apparent that considerable effort and resources are being devoted to a variety of monitoring schemes.

In the course of the preceding descriptions of national monitoring schemes a number of issues concerning monitoring methodologies, data handling and analysis techniques have emerged. The principle points to be drawn from the experience of these surveys are summarised below:

(a) Monitoring programmes need to be devised in the context of clearly defined aims and objectives, taking full account of the costs and limitations of the techniques available.

(b) All methods and systems are liable to some degree of inaccuracy. Errors are inevitable and it is therefore essential to identify the source and estimate the size of these errors. Furthermore, it is important to ensure that any conclusions drawn about recorded changes are valid within the margins of error identified.

(c) Accuracy assessment involves the comparison of features or land use classes 'predicted' from air photo-interpretation with 'known' features identified by field inspection. Although often assumed to be definitive, ground data are also liable to errors and in some circumstances can be less accurate than the predicted data.

(d) Aerial photography is currently the only proven operational remote-sensing technique. Although satellite data may provide general contextual information about change, it is unlikely to be a primary data source in the foreseeable future.

(e) Most operational systems will require the integration of more than one type of data. Commonly aerial photography is used in conjunction with ground survey.

(f) Digital cartographic techniques and the development of geographic information systems are of increasing importance, reducing measurement errors and enabling separate data sets to be overlaid and compared.

(g) The methods and operational definitions used for surveys need to be fully documented to facilitate subsequent re-use of the data by different personnel.

(h) The duplication of surveys is not necessarily undesirable, especially when techniques are still being developed. Useful knowledge or ideas can be gained by comparing different surveys even when there is a contradiction in the results.

References

Barr, C.J., Benefield, C., Bunce,R.G.H., Riddsdale, H. and Whittaker, H.A. 1986. *Landscape Changes in Britain*. Institute of Terrestrial Ecology, Monks Wood, Huntingdon.

Bunce, R.G.H., Barr, C.J. and Whittaker, H.A. 1981. *An Integrated System of Land Classification*. Annual Report of The Institute of Terrestrial Ecology, 1980, pp. 28–33.

Coleman, A. and Maggs, K.R.A. 1968. *Land Use Survey Handbook* (5th edn). Isle of Thanet Geographical Association.

Fuller, R.M., Brown, N.J. and Mountford, M.D. 1986. Taking stock of changing Broadland, I. Air photo-interpretation and digital cartography. *Journal of Biogeography*, 13, 313–26.

Hunting Surveys and Consultants Ltd. 1986. *Monitoring Landscape Change, Volume I*. Department of the Environment (London). The Countryside Commission (Cheltenham).

Nature Conservancy Council. 1987. *Changes in the Cumbrian Country-side*. NCC. Peterborough.

Stamp, L.D. 1962. *The Land of Britain: Its Use and Misuse*, 3rd edn. Longman, London.

Appendix 2.A ITE land classification

Descriptions of main classes
  1. Undulating country, varied agriculture, mainly grassland
  2. Open, gentle slopes, often lowland, varied agriculture
  3. Flat arable land, mainly cereals, little native vegetation
  4. Flat, intensive agriculture, otherwise mainly built-up
  5. Lowland, somewhat enclosed land, varied agriculture and vegetation
  6. Gently rolling enclosed country, mainly fertile pastures
  7. Coastal with variable morphology and vegetation
  8. Coastal, often estuarine, mainly pasture, otherwise built-up
  9. Fairly flat, open intensive agriculture, often built-up
 10. Flat plains with intensive farming, often arable/grass mixtures
 11. Rich alluvial plains, mainly open with arable or pasture
 12. Very fertile coastal plains with very productive crops
 13. Somewhat variable land forms, mainly flat, heterogeneous land use
 14. Level coastal plains with arable, otherwise often urbanised
 15. Valley bottoms with mixed agriculture, predominantly pastural
 16. Undulating lowlands, variable agriculture and native vegetation
 17. Rounded intermediate slopes, mainly improvable permanent pasture
 18. Rounded hills, some steep slopes, varied moorlands
 19. Smooth hills, mainly heather moors, often afforested
 20. Midvalley slopes, wide range of vegetation types
 21. Upper valley slopes, mainly covered with bogs
 22. Margins of high mountains, moorlands, often afforested
 23. High mountain summits, with well drained moorlands
 24. Upper, steep, mountain slopes, usually bog covered
 25. Lowlands with variable land use, mainly arable
 26. Fertile lowlands with intensive agriculture
 27. Fertile lowland margins with mixed agriculture
 28. Varied lowland margins with heterogeneous land use
 29. Sheltered coasts with varied land use, often crofting
 30. Open coasts with low hills dominated by bogs
 31. Cold exposed coasts with variable land use and crofting
 32. Bleak undulating surfaces mainly covered with bogs

Appendix 2.B: NCC National Countryside Monitoring Scheme

List of features to be recorded

Group A	Group B
*Hedgerow (including hedgerow with trees)	Hedgerow without trees Treeline, including hedgerow with trees
Woodland	*Semi-natural broadleaved woodland Broadleaved plantation Semi-natural coniferous woodland *Coniferous plantation Semi-natural mixed woodland Mixed plantation Young plantation Recently felled woodland
Parkland	
*Scrub	Scrub, tall Scrub, low
Bracken	
*Heathland	Dwarf shrub heath, lowland *Dwarf shrub heath, moorland
Mire	*Blanket mire Lowland raised mire
Wet ground	
Marginal inundation	
*Open water	Standing natural water Standing man-made water Running natural water Running canalised water
Grassland	*Unimproved grassland Semi-improved grassland *Improved grassland
*Arable	
*Bare rock and soil	Unquarried inland cliff and outcrop Quarries and open-cast mines, including spoil Other bare ground
*Built land	

*Habitats where 10 per cent net change is estimated with 95 per cent confidence.

Appendix 2.C: full classification of features used for the monitoring Landscape Change Project

A. *Linear Features*
  - A1   Hedgerows
  - A2   Fences and insubstantial field boundaries
  - A3   Walls
  - A4   Banks with or without low hedges
  - A5   Open ditches
  - A6   Woodland Fringe
  - A7   Urban boundary

B. *Small or Isolated Features*
  Note: These categories to be analysed in relation to data from Forestry Commission sample strips only

  - B1   Isolated trees in hedgerows
  - B2   Isolated trees outside hedgerows
  - B3   Group of trees, mainly broadleaved (< 1/4 ha)
  - B4   Group of trees, mainly coniferous (< 1/4 ha)
  - B5   Linear features (strips of woody vegetation < 20 m width and < 25 m length)
  - B6   Farmland ponds

C. *Woodland*
  - C1   Broadleaved high forest
  - C2   Coniferous high forest
  - C3   Mixed high forest (intimate mixture)
  - C4   Scrub (defined during fieldwork only)

D. *Semi-Natural Vegetation*

D1	Upland heath	(a)	—	ling (*Calluna*)
			—	bell heather (*Erica*)
		(b)	—	bilberry (*Vaccinium*); subsequently amalgamated with Dla
D2	Upland grass moor	(a)	—	smooth grassland; fescues/ bents (*Festuca, Agrostis*)
		(b)	—	coarse grassland; purple moor grass (*Molinia caerulea*)
		(c)	—	mat grass (*Nardus stricta*); subsequently

| | | (d) | — | amalgamated with D2b blanket bog; cotton 'grasses (*Eriophorum spp*), peat-forming bog mosses (*Sphagnum spp*) |

D3   Bracken

D4   Lowland heath   (a) — rough grassland
                                    (b) — heather
D5   Gorse

E. *Farmed land*
    E1   Cultivated land   (a) — ploughed/cropped land;
                                         cereals, leygrasses, legumes
                                         (crops identified during
                                         fieldwork only)

          Market gardens   (b) — including glass houses,
                                         nurseries and soft
                                         fruit farms
                                    (c) — orchards
                                    (d) — hops; subsequently
                                         amalgamated with E1c
    E2   Grassland   (a) — improved pasture
                                    (b) — rough pasture
                                    (c) — neglected pasture

F. *Water and Wet Lands*
    F1   Open water   — coastal or estuarine
    F2   Open water   — inland (not rivers)
    F3   Wetland vegetation   (a) — peat bog (valley, raised, moss)
                                    (b) — freshwater marsh (reed swamp)
                                    (c) — saltmarsh

G. *Other land*
    G1   Non-vegetated peat   subsequently amalgamated
                                         with F3a
    G2   Bare rock
    G3   Sand   — dunes, dune-slack, shingle
    G4   Developed land   (a) — built-up land; housing
                                         (including gardens), industrial,
                                         agricultural
                                    (b) — urban open space; sports fields.
                                         parks, cemeteries
                                    (c) — transport, routes
                                    (d) — quarries, mineral workings
                                    (e) — derelict land; abandoned
                                         industrial sites and
                                         mineral workings

Appendix 2.D: Monitoring Landscape Change Project

Area features: summarised results for the whole of England and Wales, 1947–1980

Feature	1947 Cover per cent	Relative standard error* %	1969 Cover per cent	Relative standard error* %	1980 Cover per cent	Relative standard error* %
Broadleaf	5.6	5.7	4.7	5.7	4.2	5.5
Coniferous	0.7	24.0	2.2	19.8	2.7	16.5
Mixed	0.7	14.6	1.0	12.0	0.9	11.7
Woodland	7.0†	5.3	7.9†	6.8	7.9†	6.8
Upland heath	3.0	20.3	2.4	26.2	2.4	25.5
Upland grass (smooth)	1.2	20.8	0.9	24.8	0.6	28.0
Upland grass (coarse)	4.6	14.6	4.3	17.5	3.9	18.8
Blanket bog	0.7	27.7	0.7	27.4	0.7	28.5
Bracken	1.1	20.4	1.0	23.4	1.0	25.2
Lowland grass heath	1.5	16.0	0.4	24.3	0.3	29.5
Lowland heather	0.4	22.8	0.2	34.5	0.2	31.6
Semi-natural vegetation	12.6†	7.4	10.1†	9.6	9.2†	10.2
Cultivated land	28.1	2.7	31.7	2.5	35.4	2.6
Improved grassland	38.1	2.8	34.5	3.1	31.0	2.9
Rough grassland	2.9	9.0	2.2	8.7	2.2	9.2
Neglected grassland	3.7	8.8	3.6	8.1	3.1	9.7
Farmed land	72.7†	1.4	72.1†	1.4	71.8†	1.4
Water/Wetland	1.3†	18.3	1.1†	17.2	1.1†	16.7
Built-up land	4.5	5.8	6.5	5.2	7.3	5.1
Urban open space	0.7	13.5	1.1	10.2	1.3	9.6
Transport routes	0.5	19.1	0.5	14.3	0.5	11.7
Other land	6.4†	5.5	8.8†	5.0	10.0†	4.9
Total	100.0		100.0		100.0	100.0

*Relative standard error is equal to coefficient of variation.
†Figures for sub-totals include rare features for which percentages are not presented in this summary table. Full tables are presented in Volumes 3, 4 and 5 of the Final Report.

National Reports

3 Belgium
A. Noirfalise Faculté des Sciences Agronomiques, Gembloux

Agricultural grassland

Distribution and landscape

Agricultural grassland is very extensive in Belgium, covering 668,559 ha
(Table 3.1) or 47 per cent of all agricultural land, which itself represents

Table 3.1 Extent of agricultural grassland (1985)

Agricultural regions	Total grassland (ha)	Per cent of total agricultural land	Permanent pasture	Mixed meadow*	Temporary meadows
			Percentage of grassland		
A. *Crop farming dominant*					
Polders	25,484	37.8	68.9	23.9	7.2
Sand Loamy region	95,696	35.6	65.5	32.5	5.0
Loess region	76,341	22.9	80.0	17.8	2.2
Condroz	56,097	41.5	68.0	30.3	1.7
B. *Mixed farming Lowlands*					
Flemish Sandy region	89,935	51.6	58.0	32.0	10.0
Campine Sandy region	64,758	65.2	38.3	51.8	9.9
C. *Mixed farming Uplands*					
Famenne	46,301	70.8	55.2	43.4	1.4
Fagne	12,781	82.2	55.5	43.7	0.8
Jurassic region	24,513	74.9	49.2	49.1	1.7
Ardenne	88,028	81.4	45.0	53.6	2.5
D. *Grassland farming*					
Eastern Belgian region	59,012	91.2	38.5	61.3	0.2
Eastern High Ardenne	29,653	97.7	29.0	71.0	—

*Hay plus grazing

Faculté des Sciences Agronomiques, 5800 Gembloux, Belgium.

less than 50 per cent of the national area. Grassland is unequally distributed through the country. It increased from 1950 to 1970, but then decreased by about 18 per cent till now, partly by ploughing for silage maize or other cereals, partly by allocation to non-agricultural use.

In the areas of region A (crop farming dominant), grassland is confined to the heavy and wetter soils and belongs to an open field landscape with scattered trees (poplars, tadpole willows). In the areas of region B, grassland is to be found more in the valleys with rows of poplars and willows, except in the grassland recently established in Campine on heathland clearings. In the areas of region C, agricultural grassland is more extensive, with many hedgerows and trees and significant areas of woodland. The areas of region D are typical *bocage* landscapes with a rather dense network of hedges, which dates back to about 1750 in the Pays de Herve. Old hedges are formed by hawthorn, hornbeam, oak, ash and holly; to which, in the limestone areas may be added, common maple, elm, gooseberry and barberry. In the High Ardennes, beech and sycamore are the main components of the old hedges. These regions are lovely landscapes that are good habitats for wildlife.

The old semi-natural grassland and hay meadows on acid soils (*Nardus*, *Molinium*) or limestone (*Bromus*) are nowadays only present in nature reserves. Moreover, intensive management has decreased the floristic diversity of many meadows since 1950. It must also be added that pasture commons no longer exist in Belgium.

Grassland management

Most grassland is managed as permanent pasture land (see Table 3.1). Because of the heavy input of mineral and organic nitrogen fertilisers in the sandy lowlands, it must be resown about every five years to prevent weed development. In the uplands, pasture is rarely renewed and white clover is frequent. In these regions, use of nitrogen fertiliser varies from 100 to 200 kgN/ha, but in the lowlands from 200–400 kg or more are used on sandy soils.

Animal grazing densities in pastures during the growing season are generally between 3 and 5 livestock units per ha, according to the amount of fertiliser used. Rotation of grazing is a general practice; permanent grazing is less productive, unless more nitrogen is given. Grass silage in spring or maize silage in autumn are also common practices, depending on the region.

Herbicides are not necessary when management is good, notably in meadows with white clover that are used more for meat production than for milk production. Other biocides are occasionally used in damp areas to control fluke-worm and *Tipula*, which caused significant damage in 1986 and 1987.

Wildlife and grassland

Hedges are very beneficial to wildlife conservation and farmers have no objection to maintaining existing hedgerows, especially in the rotation grazing system. However, from, 1770 to 1960 hedgerows decreased by 75 to 80 per cent in the areas of region D. Continuing losses are often the consequence of plot rationalisation carried out with FEOGA subsidies. These operations are preceded by a study of the ecological biotopes and amenity features, but attention given to the resulting recommendations depends on the good sense and ability of the people in charge of the rationalisation plan. Ten years ago, grassland parcels enclosed with hedges were given a negative value in the calculations preceding rationalisation. This was the primary cause of hedge removal by farmers.

Other causes of the loss of hedges include cutting for fire wood and legal measures against fire blight on hawthorn, which is a source of the disease affecting fruit trees and fruit-tree nurseries.

Encouraging management beneficial to wildlife

Generally, farmers consider that milk quotas cannot be compensated for by less intensive grassland management, because of the high rent that must be paid for grassland. They prefer to withdraw land of low quality from production. Reafforestation would be possible but income is problematic because of the high estate duties on inheritance. Alternative land use (tree planting, housing, recreation, nature protection) is already regulated by the legally binding structure plans for the countryside.

The management of hedges demands both labour and money. There is a general desire that some public action should be taken for the conservation or management of old and historically significant hedgerows in grassland landscapes.

Wetlands

Semi-natural wetlands included in farm holdings were used before 1950 as hay grassland. These were located for the most part in the large west valleys of the lowlands (Flanders, Campine, Brabent) and to a far lesser extent in the narrow valleys of the uplands, which generally have acid meadows with *Molinia* or richer alluvial meadows, all with interesting flora and fauna. There is no information about their actual area in the mixed landscapes which comprise both drained and undrained areas.

Since 1960, improvement by drainage, undertaken with the FEOGA funding, has been about 1,000 to 1,200 ha each year, mainly in North

Belgium (86 per cent). Other areas have been afforrested by poplars. In the uplands, mainly the Ardennes, conifers have been planted for afforestation, but many little parcels of land have been abandoned, and so returned to wild grassland.

Nature reserves, created either by purchase, or more often by agreement with the proprietor, take a small area of the agricultural wetlands (about 3,000 ha). The remaining wetlands in agricultural landscapes undergo eutrophication and some contamination by pesticides and, in grazing meadows, by herbicides used to control infestations of the fluke-worm, as already mentioned.

Reclamation of wetlands for productive agriculture is now of very minor importance and possibilities for conservation are good in the light of the new agricultural policy of the EEC.

Management of field margins

Most ancient field margins have been obliterated by the field rationalisations (*remembrements*), including hedgerows (see above). The trimming of hedgerows is still traditional but recommendations have been given by public authorities concerning the choice of plant species for the establishment or renovation of hedges. Little scientific knowledge is available about the beneficial or adverse effects of hedges. Some Belgian research has been concerned with the insect fauna and flora of hedges, and also with their importance as a reservoir for plant aphids and their predators.

Monitoring landscape and wildlife habitats

The historical evolution of quantitative changes in agricultural landscapes and habitats is now possible by comparing the maps of 1775 (1:27,000) and the present situation, as given by the national ecological mapping initiated a few years ago (1:25,000). While botanical information on wildlife habitats is extensive, zoological information is limited.

Some of this ecological information is already computerised, but there is no general programme concerning these matters: research remains an initiative of the university services and private naturalists. In the Northern Region of Belgium, the recently founded Institute for Nature Conservation is competent to undertake these tasks.

Much could be done in the next few years of the basis of biological data collected during the last decades. To this end, field ecologists should give as much interest to the agricultural landscape as they do to nature reserves and specific natural biotopes, which often have their exclusive preference.

Republic of Ireland

J. Mulqueen, Agricultural Institute,
Ballinrobe

Distribution and physical environment of grasslands and wetlands
The total area of the Republic of Ireland is 6.9 million hectares of which
62 per cent is grassland (Table 4.1). The area in arable and forestry is quite
low. The areas shown under rough grazings, forest and other include
most of the areas and places which have high priority for conservation.
They include boglands, oakwoods, landscapes of outstanding natural
beauty such as the Burren Karstland and wetlands such as turloughs
(seasonal lakes).

Table 4.1 Land use in the Republic of Ireland

Area	Pasture	Hay/ silage	Rough grazing	Arable	Forest	Other*	Total
m.ha	3.0	1.3	1.0	0.5	0.4	0.8	6.9
%	43	19	14	7	5	12	100

*Includes bogs, rocky land, water, urban areas

Table 4.2 shows the distribution of major land classes in the Republic
of Ireland. Most of the agricultural production comes from grasslands
on dry lowland mineral soil, which amount to 46 per cent of the land
area. The lowland and drumlin wet mineral soil grasslands comprise
32 per cent of the agricultural land in Ireland. These soils are all in
grassland and they pose particularly difficult problems to farming in
the Irish climate. They include some ecologically important wetlands
and important wildlife habitat.

The most important parameters of the Irish climate in a European
context in relation to grasslands and wetlands are the rainfall and the

evapotranspiration. The country is characterised by a surplus of rainfall over potential evapotranspiration throughout the year (Table 4.3). This surplus is greater in the west, amounting to 20–380 mm on average, increasing toward the mountains. Only a few areas exist on the east coast where on average there is a small deficit. Some summers in the west can be very wet. The range of rainfall-potential evapotranspiration balances is similar to those in the United Kingdom except that there are greater deficits in summer in the east of England.

Table 4.2 Distribution of major land classes in the Republic of Ireland

	m. ha	%
Mountain and hill, mostly > 500m	0.4	7
Hill, mostly 150–365 m and dry	0.4	7
Rolling lowland: dry mineral soils < 150m	2.6	46
Rolling lowland: wet mineral soils < 150 m	1.2	21
Drumlin soils: mostly wet	0.7	12
Peatland	0.4	7

Table 4.3 Range of rainfall and potential evapotranspiration in the Republic of Ireland (mm)

		East	West
Rainfall	Annual	700–1,200	1,000–1,600
	April–Sept.	350–500	400–650
Potential evapo-	Annual	420–470	360–390
transpiration	April–Sept.	370–410	310–340

Changes in grassland management in recent times

Livestock numbers
In recent years there have been very significant developments in grassland management. Livestock units increased almost 50 per cent from 4.4 to 6.4 millions in the 1968–80 period. This increase could have only been partly offset by a decline in arable area of 0.5 million hectares, which would allow for about one million additional livestock units. The remainder of the increase has come through increased use of fertiliser and drainage of grasslands and improved utilisation of land, such as substitution of silage for hay.

Fertilisers and animal feedstuffs

Table 4.4 Usage of commercial fertilisers in some EEC countries, 1979/1980 (kg/ha)

	Netherlands	Denmark	United Kingdom	Ireland	EEC
Straight fertilisers	260	69	44	26	72
Compound fertilisers	82	171	77	77	94
Nitrogen (N)	240	136	71	43	75
Phosphate (P)	41	46	24	27	46
Potash (K)	61	59	25	33	44

Table 4.5 Forage and concentrates as percentage of total animal feedstuffs in some EEC countries, 1979/1980

	Netherlands	Denmark	United Kingdom	West Germany	Ireland
Forage	35	32	60	48	80
Concentrates	65	68	40	52	20

While the changes have been significant, still the intensity of farming is low compared with the Netherlands, Denmark and the United Kingdom. This is illustrated in Table 4.4 in relation to fertiliser usage and in Table 4.5 in relation to usage of feedstuffs for livestock. Table 4.4 shows that Ireland uses less than the average EEC amount of all the commodities mentioned. Table 4.5 indicates that there is a large importation of concentrates into all-livestock farms in some EEC countries such as the Netherlands and Denmark. This enables higher stocking rates to be carried than would be possible from the production of forage and concentrates on the land itself. There are correspondingly larger proportions of animal manures to be disposed of, causing greater pressures on the environment in those countries. In Ireland, forage (produced on the individual farms) constitutes 80 per cent of animal feedstuffs, causing little pressures on the environment. Only in one river-lake catchment, where high pig populations are associated with grasslands on heavy clay soils of low permeability, are there significant environmental pressures. Compared with the Netherlands, Belgium, Denmark and West Germany, which have 2.87, 2.08, 1.22 and 1.25 livestock units per hectare of land, Ireland is low with only 0.89.

Conservation of grass

The Irish climate is unfavourable to hay making because of the high frequency of rainfall. Total rainfall is very highly correlated (correlation coefficient = 0.924) with the frequency of daily amounts > 2 mm. As the intensity of grassland farming increased, the risks associated with hay making and the frequent poor quality made change-over to silage essential. After a slow take-off in the 1960–69 decade from 0.3 to 3 million tons, silage production has increased almost linearly in the 13 years to 1982 to 14 million tons. This change-over has resulted in increased amounts of silage effluent of very high BOD. This must be carefully collected and either disposed of on land or fed to livestock.

Regional changes in grassland management

Intensification of grassland has been greatest in the dry mineral soils which have only minimum constraints. The least response in intensification has been on heavy soils with clay layers of low permeability. Carrying capacity increases from 1 livestock unit (LU)/2 ha on very wet land to 2.t LU/ha on free draining land. Heavy clay soils have low productivity and slow growth in spring creating an imbalance in seasonal production. Their very low bearing capacity when wet is an even greater problem giving rise to much treading damage by livestock and problems with machinery. Little or no changes in grassland management have taken place on very wet or seasonally ponded soils in some areas. There has also been less intensification of grassland management in the climatically wetter west than in the drier east.

Environmental aspects of changes in management of grassland

Changes in fertiliser usage

To achieve the increase in livestock numbers and their enhanced productivity, it has been necessary to raise grassland productivity through increased use of fertilisers. From an environmental point of view, phosphorus (P) and nitrogen (N) are the most significant if there is much run-off or leaching since they promote eutrophication of surface water bodies and nitrogen can give rise to dangerously high levels of nitrates (NO_3) in groundwaters (see Table 4.6). In the case of phosphorus, usage increased slowly from 35,000 tons to 50,000 tons over the 1958–67 decade. From 1967 to 1974 usage increased steadily to 90,000 tons. It fell off in 1975 to about 50,000 tons after a wet season and low livestock prices. After this it increased again to about 80,000 tons in 1978. Nitrogen usage has increased curvilinearly. From 1958 to

Table 4.6 Nitrate concentration in sixty-seven south-eastern rivers and sixty-eight western rivers in the Republic of Ireland, 1983

	South-eastern rivers	Western rivers
Median values (mg/l)		
Range	0.3–5.2	01.–1.7
Median	2.3	0.6
Maximum values (mg/l)		
Range	0.7–10.8	0.1–3.0
Median	3.0	1.3

Note: WHO upper limit = 11.3 mg/l

1966 usage increased only from 20,000 tons to 30,000 tons. Afterwards, it increased linearly from 30,000 to 220,000 tons in 1978. This corresponded with greatly increased grass production on the more intensive farms.

*Effect of increased use of fertilisers on soil
and water quality*

There has been a steady increase in the levels of soil phosphorus content extractable with Morgan's reagent from 0.5 to 6.0 ppm P from 1954 to 1973. From 1973 to 1980 levels have levelled out asymptotically at about 6.5 ppm P.

Mean nitrate levels (Table 4.6) are well below the WHO upper limit of 11.3 mg/l. They are much higher in south-eastern rivers, where intensive grassland farming is more common, and lower in the west, where intensive grassland farming is less common and rainfall (Table 4.3), and thus dilution, much higher.

Experimental results on nitrate leaching into drains and groundwater and on nitrate levels in surface run-off waters are available (Table 4.7). These show that at the levels of fertiliser use at present, namely less than 200 kgN/ha, high nitrate levels are not likely in groundwater. When nitrate fertilisers are applied to dry soil in the growing season, there is little leaching into drains. The dangers are in leaching/run-off of nitrates and ammonia from fertilisers and animal slurries if these are applied on wet soil. Management specifications and guidelines are now

Table 4.7 Fertiliser inputs and losses in drainage water on a heavy clay, 1966–71 in kg/ha

	Ca	P	NO$_3$N	K
Total applied 1966–71	6,405	445	828	1,295
Lost in surface run-off				
(undrained) plot 1967–71*	451	11	26	55
Lost in drainage, 1967–71**	805	2	74	23

*Total run-off = 1.9 m. **Total drainage = 3.0 m. 1967–71)
Source: Burke, W., Mulqueen, J. and Butler, P. 1974. *Irish Journal of Agricultural Research, 13,* 203–14.

available for the application of slurries to grassland.

Conclusions and practical possibilities of changes in grassland management in relation to environmental impact

The previous sections illustrate that substantial intensification in the use of grasslands has taken place. Most of this has been on dry mineral soils. Very large blocks of wet grasslands and peatlands as well as wetland and dry rocky enclaves exist, which have practically been untouched by intensificaiton. Intensification has been achieved in harmony with the environment with few exceptions up until now. This is in contrast with the very intensive conditions prevailing in countries like the Netherlands, Denmark, West Germany and the United Kingdom, with considerable pressures on the environment. Still, the situation in Ireland needs to be kept under review.

Reductions in usage of nitrogen and phosphorus are the two main practical possibilities to further reduce the impact of intensive management of grassland on the environment. It has been shown that white clover can be successfully managed to fix up to 200 kgN/ha on intensively used pastures. A stocking rate of 2.5 cows/ha can be carried. There are problems of shortage of feed in dry and wet summers as clover fixation of nitrogen is adversely affected and standby supplies of fodder and concentrates are required for these periods.

There is evidence that the use of phosphates is tending to level off. Some experimental results indicate that intensively fertilised land may have enough soil phosphorus for optimum growth on grazing land. The application of phosphates could be reduced and even stopped completely for periods in some of these lands. Other issues on grassland in relation to the environment can be dealt with by good management in accordance with available guidelines and specifications. In particular, the level of usage of biocides in Irish grasslands is practically non-existent and levels in surface waters are essentially the background levels.

Wetlands

Wetlands in Ireland are chiefly of importance as wildlife habitat and a small proportion are important to plant and animal communities or constitute unique landscape. Practically all wetlands are used by farmers for extensive livestock grazing. Table 4.2 shows that about half of Ireland consists of wetland. Most of this consists of wet mineral soils in grassland (33 per cent of Ireland) with another 7 per cent in peatland. From 1800 onwards practically all the wet mineral grasslands have been drained at one time or another, some under famine relief schemes. This drainage was often only partly successful because of poor outfalls, low permeability soils or because the geohydrology was not understood. In recent times, there is improved knowledge of the geohydrology for some important areas but there exist many complex areas where little is known and Irish clay soils of low permeability require much more research to understand their behaviour. The information already researched is applicable to and essential for nature management.

Wetlands in Ireland may be classified as wet grassland, fen, raised and blanket bogs, seasonal lakes (turloughs), lake edges and river flood plains and coastal wetlands. There are 90,000 and 560,000 ha of virgin raised and blanket bog respectively and 92,000 ha of man-modified fen peat. Little has been done by way of drainage on the large flood plain catchments of the River Shannon complex. Only five small flood plain areas have been drained in recent times. The total area of flooded land in the Robe–Mask, Boyle and Bonet Rivers is 30,000 ha of which 23,000 ha have benefitted. The other 7,000 ha remains wet. Only a few turloughs have been drained and some small lakes such as Island Lake near Ballyhaunis have been converted from a lake to a wetland with better feeding grounds for wildlife. While there has been much talk about the impact of drainage on the environment, the drainage has been planned for the most part in a way sympathic to nature issues.

Under the Western Drainage Scheme a total of 162,000 ha of wet grassland has been drained. This is practically all wet grassland where old drainage works have fallen into disrepair or were not effective due to too distant spacing or incorrect design in the nineteenth and early twentieth century. Little new land has been brought into production under this scheme. The total drained is only 8.5 per cent of the total wet grassland. Lake edges, river flood plains, coastal wetlands, swamp woodlands or turloughs were not drained under this scheme. The drainage and redrainage of wet grassland sections is essential for many farmers so that they can put together a viable farm. Failure to do this can only result in an outward migration of farm families and significant change in the landscape. Irish farmers are very sympathetic to nature issues and conservation. The healthy state of nature in Ireland is indicative of this at the present time. Ireland is well endowed with virgin wetlands, to which little or no danger would be predicted before the end of the century.

References

Cabot, D. ed. 1985. *The State of the Environment*. An Foras Forbartha, Dublin.

Gardiner, M.J. and Radford, T. 1980. *Soil Associations of Ireland and their Land Use Potential*. An Foras Taluntais, Dublin.

Hammond, R.F. 1979. *The Peatlands of Ireland*. An Foras Taluntais, Dublin.

Lee, J. Aspects of agricultural land use in Ireland. *Environmental Geological Water Sciences*, 9, 23–30.

Mulqueen, J. Hydrology and drainage of peatland. *Environmental Geological Water Sciences*, 9, 15–22.

Robinson, M., Mulqueen, J. and Burke, W. 1987. On flows from a clay soil: seasonal changes and the effect of mole drainage. *Journal of Hydrology*, in press.

Sherwood, M. 1986 Impact of agriculture on surface water in Ireland. Part II, prospects for the future. Environmental Geological Water Sciences, 9, 3–10.

Toner, P.F. 19886. Impact of agriculture on surface water in Ireland. Part I general. *Environmental Geological Water Sciences*, 9, 3–10.

5 | Luxembourg

J. Frisch, Administration des Services
Techniques de L'Agriculture, Luxembourg

Management of grassland

Historic changes in use and management of agricultural grassland

Agriculture in Luxembourg has been marked in the twentieth century by a profound modification of the grassland/arable land ratio. Since the beginning of the century, there has been a continuous expansion of forage crops at the expense of arable crops, especially wheat; the importance of this expansion, which is due mainly to permanent pastures and meadows, is illustrated in Table 5.1. The proportion of land devoted to permanent grass has increased from 17 per cent in 1900 to 56 per cent in 1986, and these figures do not include the leys (8 per cent), which are included in the 'arable lands'. The total area of permanent grassland in Luxembourg is 70,600 ha.

Table 5.1 Increases in forage crops from 1900 to 1980 (as percentage of total agricultural area)

	1900	1920	1940	1960	1980	1986
Permanent hay meadows	16	16	17	19	22	232
Permanent pastures, grazed	1	7	12	27	32	33
Total permanent grassland	17	23	29	46	54	56
Arable land	80	73	68	52	44	43

The overall examination of forage production in Luxembourg in the last decades shows the following facts:

1. There was a spectacular increase of permanent grazed pasture, from 1 per cent in 1900 to 33 per cent in 1986.
2. There was a very small increase of permanent hay-meadows. This is due to the fact that the hay-meadows cover land which cannot be tilled, and are the most profitable form of utilization of such areas.
3. A quite noteworthy decrease of pure stands of Red Clover and Lucerne occurred. This decrease is compensated by a simultaneous increase of the areas covered by leys, the latter being made up of mixtures of grasses and clovers (Kleegras) and representing about 8 per cent of the agricultural area.
4. The very strong decrease, or even disappearance, of fodder-beets was completely compensated for by the spectacular extension of silage-maize.
5. As a complement to this extension of forage production, a number of more rational harvesting and grass conservation techniques have developed, the most important of which is silage making.

The profound transformation to which agriculture in Luxembourg has been subjected during the twentieth century is closely linked to the economic and social evolution as well as to the natural conditions of the country.

(a) Among the economic and social reasons that were responsible for the considerable increase of the forage area, the continuous growth of meat and milk requirements of a highly industrialized country are most important. In our markets, there is a larger absorption of animal products; the consumption of plant products (potatoes, starchy products) is decreasing, so that agriculture in Luxembourg is oriented more and more towards the production of meat, and the land devoted to the growing of grass covers more and more ground. The production of beef and milk alone constituted 90 per cent of the total agricultural production in 1983, as compared to only 53 per cent in 1956 and 38 per cent in 1935.

(b) The structure of agricultural holdings in Luxembourg is characterized by the predominance of small-sized or medium-sized family farms (25–35 ha). The development of cattle-rearing could rather easily become integrated into the internal structure of Luxembourg farms, for it is cattle, and above all milk production, which makes it possible to make best use of family labour and get the best profit out of existing farm buildings.

(c) Luxembourg is a country with rather poor soils and a wet and irregular climate (see Figure 5.1), and is thus better suited for cattle-rearing than for cereal growing. The areas devoted to bread cereals have very much decreased since 1967, the first year of application of Common Market prices, which were much lower than the former national prices. From that time it was clear that the role of agriculture in Luxembourg could only be found in livestock production, milk and beef. Agricultural land, the quality of which is not high enough to produce wheat and sugar beet in a competitive way for the Common Market, are best utilized for forage; thus farmland was oriented quite naturally towards herbage production.

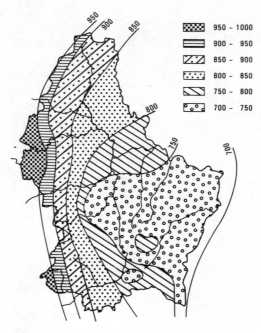

▓▓▓▓	950 – 1000
▭▭▭	900 – 950
▨▨	850 – 900
⠐⠂⠄	800 – 850
◚◚	750 – 800
⦿⦿	700 – 750

Figure 5.1 Annual rainfall in Luxembourg (mm)

Structure of grassland production in Luxembourg

From the point of view of forage production the Grand Duchy of Luxembourg may be divided into two distinct natural regions differing both by the nature of soil and topography, and by the climate and general agricultural system: Oesling in the north and Gutland in the south.

Oesling covers the part of the country corresponding to the Ardennes (Devonian) (see Figure 5.2); it is a high table-land, cut by narrow and deep river valleys. As these are very narrow, there are few alluvial soils in the Oesling. For this reason natural grassland and hay-meadows are rare (see Figures 5.3 and 5.4), and the farmers have

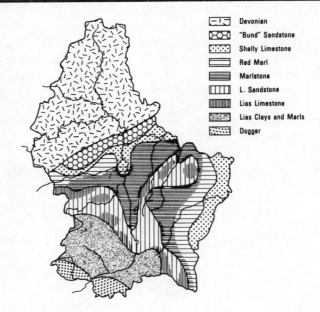

Devonian
"Bund" Sandstone
Shelly Limestone
Red Marl
Marlstone
L. Sandstone
Lias Limestone
Lias Clays and Marls
Dogger

Figure 5.2 Geology of Luxembourg

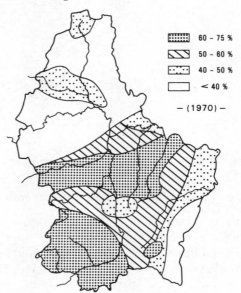

60 – 75 %
50 – 60 %
40 – 50 %
< 40 %

– (1970) –

Figure 5.3 Percentage of agricultural area in Luxembourg under permanent grassland

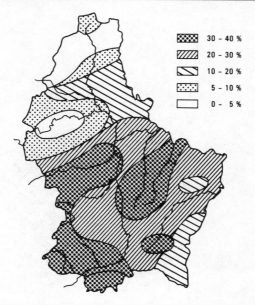

▨	30 – 40 %
▥	20 – 30 %
◩	10 – 20 %
⬚	5 – 10 %
☐	0 – 5 %

Figure 5.4 Percentage of agricultural area in Luxembourg under permanent hay meadows

to rely on the arable land of the hills for the necessary herbage. The soils of the Oesling table-land are well-drained, very permeable and generally very shallow, and therefore the grasslands often suffer from the summer drought and their degradation is fast. The farmers in the Ardennes have therefore to reseed their pastures time and again or to include them in the rotation of arable crops, in other words to create leys. The latter cover about 20 per cent of the Oesling agricultural area, and constitute the basis of forage production. In the north-west of the Oesling the climate is wetter and the soils are deeper, so that the grassland conditions are excellent in this area, specially for pastures (Figure 5.5 and 5.6)

The soils of *Gutland* are very heterogeneous and belong partly to heavy marl and clay formations, partly to light sandy formations. The heavy marl and clay soils are remarkably well suited to herbage production. This is a traditional livestock-rearing region, based on the utilization by grass of the larger part of the clay soils, which are too heavy for cereal crops. Permanent grassland covers, according to the district, from 60 to 70 per cent of the agricultural area; it constitutes the basis of the regional forage production (see Figure 5.3). The proportion of permanent grassland increases from east to west as shown in Figure 5.3, which is mainly due to the increasing rainfall in the west (cf. Figure 5.1). The determining factors for the improvement of grassland yield are fertilization and improvement

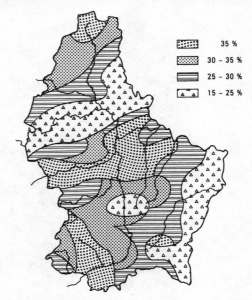

35 %
30 – 35 %
25 – 30 %
15 – 25 %

Figure 5.5 Percentage of agricultural area in Luxembourg under permanent grazing pasture

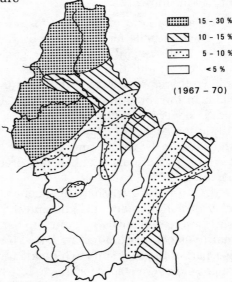

15 – 30 %
10 – 15 %
5 – 10 %
< 5 %

(1967 – 70)

Figure 5.6 Percentage of agricultural area in Luxembourg under arable fodder cropping

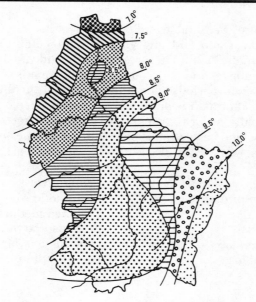

Figure 5.7 Average annual temperature in Luxembourg

of soil drainage, as there is often an excess of water.

Grassland map

In order to improve the permanent pastures and to assist better grassland management, the Ministry of Agriculture has given the Centre for Phytosociological Cartography of the Agricultural Faculty of Gembloux the task of drawing a vegetation map of the main grassland regions of the country, in collaboration with the office of agricultural technical services. This grassland map represents an ecological survey of grasslands, which makes it possible to orient the forage policy of the region according to a better knowledge of conditions and ecological restrictions, and to indicate the type of improvements to be carried out and of the total production potential. The data given by the map on grassland potential gives the farmers the possibility of applying cultivation methods better suited to the ecological circumstances, such as the improvement of soil drainage, more rational fertilizer use, and the introduction of more intensive forage crops. The vegetation map is nearly finished and covers practically the whole area of the Grand Duchy of Luxembourg.

The grassland map defines six *suitability classes* or *herbage zones* that are mainly based on the soil moisture regime, depending largely on the soil structure, water table and climatic factors.

Zone B: Dry grassland and low yield potential due to general water deficiency in the soil. The grass growth and grazing seasons are confined to spring and early summer; this zone is very sensitive to summer drought periods. The productivity of the aftermath is very low and the aftermath is normally used as extensive pasture. As the soil is normally fit for tillage, fodder growing is recommended. The natural grassland vegetation is characterised by the abundance of the drought-resistent MESO-BROMION species such as:

Bromus erectus

Brachypodium pinnatum

Ranunculus bulbosus

Salvia pratensis

Scabiosa columbaria

Centaurea scabiosa

Knautia arvensis

Primula veris

Pimpinella saxifraga

Plantago media

Sanguisorba minor

Cirsium acaule

Anonis spinosa

Zone b: Rather high yield potential due to good soil moisture regime, good soil structure and the depth of soil.

The b-zone is very well suited for pasture ('fattening pastures'), with only minor limitations: limited summer growth in dry years, and generally limited autumn growth. They are the most widespread grazing pastures in the country, and have generally developed from sown pastures originating from arable land; the soil is most suitable for cultivation or forage growing. The dominant herbage species belong to the ARRHENATHERION (hay meadow):

Arrhenatherum elatius

Anthriscus sylvestris

Campanula patula

Crepis biennis

Gallium mollugo

Geranium pratense

Heracleum sphondylium

Pastinaca sativa

Pimpinella major

Tragopogon pratensis

Lathyrus pratensis

Centaurea jacea

If this grassland is permanently grazed (pastures) the following species become dominant (LOLIETO-CYNOSURETUM):

Cynosurus cristatus

Phleum pratense

Lolium perenne

Capsella bursa-pastoris

Bromus mollis

Bellis perennis

Poa pratensis
Trifolium repens
Plantago major
Cirsium arvense

Potentilla anserina
Veronica serpyllifolia
Agropyron repens

Zone T: Highest yield potential due to stable and optimal moisture regime throughout the grazing season. They are best suited for grazing pastures. Poaching risk is low, and high stock densities are readily sustained throughout the season.
   The growth conditions for Lolium perenne and Trifolium repens are optimal (LOLIETO-CYNOSURETUM TYPICUM) and neither the drought resistant species nor the typical moist grassland species are represented.

Zone Lb: Unstable moisture regime: wet soil conditions in the spring, dry soil conditions in the summer and autumn, normally situated in the valleys or on slopes. They are usually cut for hay in the spring and afterward used as summer and autumn grazing for cattle. They are not fit for tillage but can be easily improved by drainage. The dominant herbage species are the same as in Zone b, with a clear tendency to the moisture loving species such as:

Deschampsia cespitosa
Molinia caerulea
Juncus effusus
Juncus conglomeratus
Cirsium palustre

Lychnis flos-cuculi
Ranunculus acris and repens
Cardamine pratensis
Sanguisorba officinalis
Silaum silaus

Zone L: Moist and fertile alluvial meadows, with a permanent high water table, usually cut as hay and occasionally summer grazed by cattle. Pastoral use is rather restricted on account of the high risk of poaching.
   These moist meadows are 'absolute grassland', in other words, they are not fit for tillage; they can be easily improved by drainage.
   The herbage vegetation is characterised by the abundance of CALTHION (BROMION RACEMOSI) such as:

Bromus racemosus
Caltha palustris
Cirsium oleraceum
Polygonum bistorta
Deschampsia cespitosa

Ranunculus repens
Angelica sylvestris
Filipendula ulmaria
Sanguisorba officinalis
Geum rivale

Senecio aquaticus
Alopecurus pratensis
Holcus lanatus
Poa trivialis
Cardamine pratensis
Lychnis flos-cuculi

Cirsium palustre
Scirpus sylvaticus
Myosotis palustris
Lotus uliginosus
Equisetum palustre

Zone H: Very wet alluvial meadows, situated normally at the bottom of the valleys,developed on alluvial soils with impermeable·lower strata underground, so that the water table is at, or near, the soil surface throughout the year. These meadows are nearly always cut for hay of very poor forage quality, often used as litter.

The low potential yields are often combined with extremely difficult trafficability and high risk of poaching; pastoral use is very restricted and few improvements are feasible.

The H-zone meadows include the following grassland communities: CARICION FUSCAE; MAGNOCARICION; PHRAGMITION; the dominant plant species are:

Equisetum palustre
Mentha aquatica
Caltha palustris
Lythrum salicaria
Cirsium palustre
Filipendula ulmaria
Carex gracilis
Carex panicea
Carex acutiformis
Carex disticha

Juncus acutiformis
Scirpus sylvaticus
Phragmites communis
Phalaris arundinacea
Galium palustre
Comarum palustre
Ranunculus flammula
Iris pseudocorus
Lotus uliginosus

Management of grassland

The grassland map is a very useful instrument in the management of grassland. Zones b and T are the best suited areas for intensive grass-production, specially for intensive grazing pasture, whereas the zones Lb and L can be easily drained and are most appropriate for haymaking. Zones B and H are less suited for grass production and they may be used for other purposes. As for żone B, it can be improved by ploughing up the natural sward and by sowing a seed mixture of drought-resistant forage legumes and grasses such as lucerne, sainfoin, and cocks-foot. If the soil is not too stony and the fields not too steep, permanent ley-farming would be the best forage crop growing system for these soils.

Extremely dry fields could be used for wildlife conservation. Zone H suffers from excessive wetness. Mechanical harvesting becomes increasingly difficult and the fodder-quality is very bad; drainage costs for this wet grassland are very high, and the results of drainage are poor.

As zone H supports much interesting wildlife, these areas are identified for wildlife conservation.

Management of wetland

A high content of water in soils due to inadequate drainage or too low a rate of infiltration can reduce herbage production in several ways:

—Wet soil is slow to warm and growth may be delayed in spring.
—Surface water kills the good grasses and legumes, and encourages sedges, reed-grasses, rushes and other weeds of low forage quality.
—The compressive strength of soils is greatly reduced as water content increases. Damage to the soil structure and destruction of the grassland sward results from poaching by stock and tractor wheel-slip.

The wetlands are essentially concentrated in the centre and the south of Luxembourg, and these wet soils are covered either by permanent hay-meadows or by forests. These wet areas are mainly due to impermeable soil formations (marl; clay) leading to a permanent high groundwater table. It is estimated that about 8,000–10,000 ha of wet land must be drained in Luxembourg.

Formerly (before the Second World War), these wet grassland areas were drained by open ditch drainage dug by hand, and the results were satisfactory. Today, as manpower has become very scarce on farms open ditch drainage has disappeared, and the wet areas pass progressively to marshy land. Today there are two alternatives to make use of these wet areas:

—drainage on a larger scale by special drainage machines, and by the help of a drainage association;
—creation of wildlife conservation areas.

It depends on the availability of agricultural land which of these solutions is chosen. The average areas of drainage of wetland in Luxembourg in the last twenty years has been about 100 ha per year.

As to the wetland turned to wildlife conservation, this has increased in recent years. At the moment there are about 100 wildlife reserves planned in our country, covering an area of several hundreds of hectares of extremely wet (or dry) land. Special attention is paid to the agricultural management of buffer zones between wildlife conservation areas and neighbouring land, including lowering of water-tables, effects of nutrient enrichment, agrochemicals and pollutants.

Management of field margins

The value of field margin habitats as reservoirs for wildlife in agricultural areas, as elements of the visual landscape and as shelter for cattle and birds is highly recognised in Luxembourg. Great efforts are made to conserve the existing field margin habitats and hedgerow shrubs, and to plan new ones. Specially in the openfield landscapes, new hedgerows have been planted in the past, and this will be encouraged even more in the future.

Monitoring landscape and wildlife habitats

In the framework of the grassland mapping the most interesting and valuable grassland and wildlife habitats have been noted and listed. As all grassland areas of Luxembourg are recorded on the grassland map at the scale 1:25,000 it will be relatively easy to control changes in landscape and wildlife habitats in the future.

6 The Netherlands, I: Research
M. Hoogerkamp, Centre for Agrobiological Research (CABO), Wageningen

As in other EC countries, agriculture in the Netherlands is currently practised in such a way that large amounts of high-quality food are produced at relatively low cost. In the European Community the production of various agricultural products now outstrips demand. This surplus production has a highly adverse affect on the EC budget. Other adverse effects of conventional agriculture are:

—pollution of the environment (soil, water, air);
—harm to wildlife (flora and fauna);
—damaging effects on the landscape.
Sometimes other aspects, too, are a cause of public concern:
—decrease in employment;
—high inlets of fossil energy;
—odour (from intensive livestock production in particular);
—decrease in soil fertility;
—animal welfare (animal housing systems);
—large-scale farming systems.

In the past agricultural research, extension and education focused on the agricultural sector as a producer; raising yields and improving the quality of products were important objectives with an eye to the security of the food supply and to guarantee the continuity of agricultural enterprises as much as possible.

Nowadays, the agricultural sector leans towards a more balanced approach in which attention is also paid to other functions of agriculture, and the interaction between agriculture and the environment, nature and

The author wishes to acknowledge the valuable assistance of Ms Lisbeth Spoor of the Translation Bureau of the Ministry of Agriculture and Fisheries for translating this chapter into English.

the landscape. Much is being done to reduce the adverse side-effects of agriculture including: research, extension, education, legislation and enforcement and private initiatives. This wide range of activities calls for a selection of research. In this chapter a number of relevant research activities will be discussed; De Boer and Reyrink in the next chapter elaborate on the policy concerning the relationship between agriculture and nature conservation in the Netherlands.

Reduction of adverse effects on nature and the environment (research into individual aspects)

In the research programmes of many Dutch agricultural, biological and other research institutes and universities, much attention is paid to quantifying, understanding and reducing (even eliminating) the adverse effects of certain products used in agriculture, such as manure and biocides.

This requires a great deal of effort, given the complexity of the problems. To minimalise the adverse side-effects for example of herbicides, research is aimed at solving the following questions:

—What are the adverse effects of herbicides and their metabolites on public health, the ecosystem, soil fertility, ground water quality and how significant are these effects?
—What happens to the products after application: in the plant, soil, water and air?
—How can weed control be achieved with the use of fewer active ingredients; for instance, by improving formulation, application techniques, or by responding more adequately to weather conditions before and during spraying?
—When and to what extent do weeds have to be killed (critical period, threshold values, population dynamics)?
—Are there alternative ways of weed control (biological, physical)?
—Can various solutions be combined in one management system (integrated system; see next section)?

Results from this research may, if attractive to the applicant (in particular concerning farm economics), be applied in practice; modern computer-aided systems may be of help. Should practical application proceed less successfully than required nationally, it may be adjusted by imposition of special regulations, e.g. the approval policy for pesticides.

Another key topic is the surplus production of manure (and of slurry in particular). Often the number of animals per area exceeds the safe density (affecting soil fertility, air and water). Especially in areas with intensive livestock production this is a serious problem. In many cases slurry production is so high that use on the farm is impossible without creating the above-mentioned adverse effects. Possible solutions are destruction, or increasing its value by processing and/or transport to farms with under production.

The storage of slurry, its transport and utilization require a great deal of research, e.g. into the risks of carrying pests, diseases and weeds, its heavy metal content and nutrient content (amount and ratio of minerals).

Research in itself is not sufficiently effective to solve the above-mentioned manure problems. In view of the seriousness of the problem, the Dutch government has adopted regulations on the output allowed (on the basis of phosphate content), transport, records of the manure produced, use on the farm, and on the periods in which slurry may be spread on the land. The maximum amount of phosphate allowed per ha will gradually be reduced. This will result in higher costs for the farmers whose livestock produce slurry and perhaps in a (considerable) reduction in the number of animals.

Integrated farming systems

Research into integrated systems can be performed at many levels: integrated weed management, integrated pest control, integrated crop management, integrated agriculture and nature management. Attempts are made to combine separate elements in an optimal system which can be used in practice; another important objective is to restrict the use of pesticides and fertilizers.

The various components of the system may be studied separately, after which they may be combined, manually or with the help of a computer. Theory and practice do not always correspond, however. In the Netherlands a more holistic approach has been opted for in addition to this approach. For this purpose two experimental farms have been set up, one on clay soil (at Nagele) and one on reclaimed peat soil (at Borgerswold), where an integrated farming system is compared to a conventional system. The setting-up of a third farm on sandy soil is under discussion. Crop protection, crop rotation and the application of fertilizers are key elements.

The first results of the farm on clay soil are promising. As concerns crop protection it has proved possible to economise on biocides. Occasional yield losses have been compensated by lower expenses (Vereijken, 1987). Continuation of the experiments will have to show whether this trend will continue and whether it applies to other types of soil as well. As concerns yields, it may be assumed that the possible expected losses will be relatively small and that with the increase in knowledge and the improvement of certain aspects of the integrated approach yields will gradually rise. The extent of the yield losses also depends on the extent to which one wishes to reduce undesired side-effects.

Alternative farming systems

The alternative farming systems practised most in the Netherlands are bio-dynamic and ecological agriculture (organic farming), albeit on a small scale. Both types produce for a specific group of consumers, who are willing to pay more for these products. Until recently these types of alternative farming systems depended almost entirely on private initiatives. In the last few years, however, the Dutch government has increasingly supported these systems in the form of, *inter alia*, research, education and extension.

On the above-mentioned experimental farm at Nagele (Research into Farming Systems—OBS) the bio-dynamic system is examined in addition to the conventional and the integrated system. Plans are designed to set up similar comparisons for horticulture, fruit growing and stock farming.

It may assumed that some current undesired side-effects of conventional agriculture will be reduced by using, *inter alia*, no or few fertilisers and biocides. But, especially if alternative systems are applied on a larger scale, yields will be markedly lower on average than in conventional agriculture. So it is expected, that the cost price of products will be higher than in conventional agriculture.

Other crops and new applications

The EC market for a number of products is saturated. New crops and new uses of surplus crops are therefore called for. An additional advantage of the introduction of new crops may be the extension of crop rotation. Although much time is devoted to this research, no breakthroughs have been achieved yet. In the Netherlands much attention is currently being paid to the Jerusalem artichoke, pea, onion and malting barley. Hemp, flax, and standing timber (osiers: 8–10 years and fast-growing forest: 10–20 years) are also considered subjects of research, as well as crops which supply raw materials for industry (alcohol, plastics and pharmaceuticals).

Simulation models

The experimental application of a great number of alternative farming systems on trial plots and experimental farms (private or government) cannot be realized, partly in view of the limited funding of research.

Therefore, models with which various scenarios can be simulated are a welcome supplement, of which De Wit, Rabbinge, Van Keulen and others have given marked examples. The use of simulation models is only possible if field research produces sufficient technical data which can be built into these models and if the most optimal systems are tested in practice.

The following are examples of such simulation models: rise in energy costs (Bakker cited by De Wit *et al.*, 1987b), forest growth (Mohren, 1987), rise of ambient CO_2 (Penning de Vries *et al.*, in press), planning of regional agricultural development (De Wit *et al.*, submitted for publication).

Agriculture with extended objectives

Within the framework of the so-called Relationship Memorandum (see following chapter by De Boer and Reyrink) the need was felt for supportive research. On the one hand this research is now done to examine the effects of restrictions on farm organization, farm management and farming results. Another object of current research is to what extent restrictions have contributed to the realization of landscape and scientific objectives. Further, the need was felt for simulation models, with the help of which compensation to be paid to the farmers concerned could be calculated.

Research is carried out by a team of researchers from various disciplines. Studies are performed at a great number of farms (about sixty) where restrictions have been imposed, and in two larger districts, studies are focused on grassland grazed by cattle (i.e. yields, floristic composition, roughage quality, economic results of the farms in question, meadow birds and landscape aspects). On the basis of the results from these and other studies simulation models (generalising business models) have been drafted with which different situations can be calculated.

Linear and small landscape elements

In the Netherlands linear and small landscape elements such as watercourses, pools, hedge banks and road verges are essential to the scenery and/or play an important role as natural values. Various organizations are working on a collection of the data known on these elements and to make it accessible to policy and practice. Gaps in knowledge will be filled. This applies not only to the intrinsic value of the elements concerned, but also to the interaction between these elements and modern agriculture (mutual influencing) and management in practice.

One problem is that modern agriculturalists are usually not interested in these elements or even consider them undesirable. But often farmers are the obvious persons to care for them. Management agreements between the government and farmers may offer a solution to this problem (see the following chapter).

Separation or integration

The countryside is used for various purposes such as agricultural production, recreation, living and transport, nature conservation and landscaping. A keypoint of discussion is whether separation or integration (spatial intertwining and/or integration of functions) will contribute to these functions, and if so in what way and to what extent. Their feasibility (e.g. according to the costs) is an important aspect. A study of this problem has been completed by the Department for Environmental Biology of the Leiden University, where Van de Weijden et al. (1987) and De Wit et al. (1987a) have devoted much attention to this problem.

References

Mohren, G.M.J. 1987. Simulation of forest growth, applied to douglas fir stands in the Netherlands. Dissertation, Agricultural University of Wageningen (March 1987).

Penning de Vries, F.W.T., van Keulen, H., van Diepen, C.A., Noy, I.G.A.M. and Goudriaan, J. (in press). Simulated yields of wheat and rice in current weather and in future weather when ambient CO_2 has doubled. (International Symposium on Climate and Food Security).

Vereijken, P. 1987. Ecologische gewasbescherming; van geintegreerde bestrijding naar geintegreerde landbouw. Landbouwkundig Tijdschrift 99, No. 5, 19–21.

Van der Weijden, W.J. et al. 1987. Bouwstenen voor een geintegreerde landbouw. Voorstudie en achtergronden WRR, V44. Staatsuitgeverij, 's-Gravenhage, The Netherlands.

Wit, C.T. de, Huisman, H. and Rabbinge, R. 1987a. Agriculture and its environment: are there other ways? Agriculture Systems, 23, 211–36.

Wit, C.T. de, van Keulen, H., Seligman, N.G. and Spharim, I. 1987b. Application of interactive multiple goal programming techniques for analysis and planning of regional agricultural development. Submitted to Agricultural Systems, June 1987.

7 | The Netherlands, II: Policy

T.F. de Boer and L.A.F. Reyrink,
Ministry of Agriculture and Fisheries,
Utrecht

Agriculture has always played a key role in rural areas. Many of the beautiful landscapes and valuable wildlife areas in the Netherlands are due to the economic conditions and land users' and land owners' interests of the past. When management of rural areas was mainly in farmers' hands there wa still some balance between agricultural productivity and wildlife and landscape conservation. The rapid modernization of agriculture in the last few decades, with its damaging side-effects, has created a tension between the interests of agriculture on the one hand and wildlife and landscape conservation on the other.

Some 70 per cent of the Netherlands consists of farmland with more than half of this area in grassland. Mixed farms have become rare. Farmers have specialized in arable farming, horticulture or animal production. The dairy industry is the most advanced sector. In addition, there is much pig fattening, poultry farming and beef production in the south and east of the country. Some indication is given on the trend in the dairy sector in Table 7.1, because grasslands are the most important areas with respect to wildlife and landscape conservation. It appears that grassland use has greatly intensified since the Second World War. Although total milk production has declined as a result of the EEC super-levy since 1984, the intensity of grassland utilisation has not declined.

Diversity of wildlife has suffered from the intensification of agriculture: the number of plant species in grasslands has declined sharply. Much of the former vegetation of field margins, such as wooded banks and hedges have disappeared. Although the west and north of the Netherlands still provide nesting places for large numbers of meadow birds, such as Lapwing, Black-tailed Godwit and Redshank, the numbers of some of these species have already fallen sharply.

Table 7.1 Trends in the dairy sector over the period 1950–1980

	1950	1980
Dairy cattle/hectare	1.1	1.8
Dry matter production/hectare/year	6,800 kg	11,000 kg
Nitrogen/hectare/year	70 kg N	300 kg N
Litres milk/hectare/year	4,200 kg	10,000 kg

The main causes of the loss of wildlife interest are drainage, high fertiliser and pesticide dressings, ploughing-up and re-seeding of grasslands, field enlargement, as well as earlier turning out of animals to graze and earlier first grass cut.

The origin of current policy on rural areas

Since the early 1970s, public pressure for policy instruments to redress the balance between agriculture and the environment has become more and more insistent. Parliament has repeatedly pointed to the decline of valuable agro-landscapes and to the plight of those who have given the landscape its present outlook (namely, the farmers).

What was the situation before 1974?

(1) In some areas it was not permitted to execute land rationalisation plans *with Government aid* because of the scientific and environmental value of those areas.
(2) More and more municipal land use zoning plans contained provisions that imposed limitations on farming. No compensation was given and these regulations were not very effective.
(3) For many years, farmland acquisition for the creation of nature reserves had been *ad hoc*. This met with farmers' resistance and led to fragmentary development of nature reserves.

In the long run, the management that is desirable in *nature reserve areas* (e.g. management aimed at protection of endangered meadow bird species or plots of botanic value) is incompatible with profitable farm management. Therefore these areas will be withdrawn from agriculture. Their management will be transferred to a nature conservation body, but until the land in these areas has been acquired, the farmers may conclude management agreements and during a transitional period will manage them to preserve their wildlife and landscape values.

In *management areas*, adapted management is sufficient to protect the existing environmental and wildlife values. Management areas remain agricultural areas, where production of food and raw materials is still one of the main objectives but protection of wildlife and landscape elements is also important.

The original target had been to acquire 100,000 ha, management of which would be transferred to a nature conservation organisation. In addition, 100,000 ha would be designated as management areas, where farmers could adapt farm management partly to the preservation and development of wildlife and landscape values. However, for budgetary reasons, the target for the time being is to designate no more than 100,000 ha of management and reserve areas altogether.

Broad designation of areas and final delimitation of plots

On the basis of a survey of valuable agro-landscapes, the Government proposed to the provincial authorities that in the first instance some 86,000 ha of cultivated land should be designated for implementation of the policy instruments of the so-called *Relatienota* (i.e. a memorandum on environmentally-sensitive areas). The consultations between the central and the provincial authorities got off to a slow start, because many provincial authorities appeared unwilling to pronounce judgement on the implementation of the memorandum at a time when they were drafting regional plans. In areas for which land development projects were being prepared or executed, designation of management and reserve areas was integrated into the development plan.

As soon as the broad designation of the areas that were to be covered by the memorandum had been completed, the final delimitations of plots had to be established. Frequent consultations with interested parties, such as the local authorities, water control boards and land development committees were of the highest importance. Information to the farmers involved on the effects of the delimitations of the plots was essential. When the final delimitations of plots were laid down, it had also to be decided whether the plots were to receive the status of a management or of a reserve area. For the interested parties this was important, because designation as a management area meant that the area in question would remain in agricultural use. Farmers in these areas could continue their holdings and develop them further. In addition, they had the option of concluding a management agreement. Lands in reserve areas were purchased in order to optimize management for wildlife and landscape conservation. A nature conservation body was made directly or indirectly responsible for their management. Here, too, management agreements could be concluded during the period prior to acquisition.

The broad designation of the 86,000 ha identified for the priority list has been completed, while preparations for the broad designation of the remaining 14,000 ha have been started. Final delimitation of nearly 41,000 ha (Table 7.2) of ecologically vulnerable areas has already taken place.

Table 7.2 Management agreements in the Netherlands, 1981–1986

		Management plans		Management Agreements				Manage- ment
Year	Deli- mited Number	Area	Nature reserves	Poten- tial area	Area agreed		Number of farmers	agree- ments plus reserves
area (ha)		(ha)		(ha)	(ha)		%	%
1981	19,723	9	5,743	500	5,243	967	18	72	26
1982	24,685	16	7,056	878	6,178	1,205	20	108	17
1983	26,413	28	15,627	1,726	13,901	2,947	21	290	30
1984	29,399	37	18,961	2,172	16,789	3,682	22	415	31
1985	34,333	48	20,769	2,904	17,876	4,661	26	611	36
1986	40,631	67	25,599	3,601	21,998	6,334	29	890	39

The drafting of management plans

If an environmentally-sensitive area has been delimitated and designated as a management or reserve area, the Provincial Committee for Land Management (PCBL), which consists of government officials and representatives of farmers' and nature conservation organizations, drafts a management plan. Interested parties in the region are closely involved in the drafting of this plan. In many cases a PCBL sub-committee is established, on which there are also regional representatives. In addition, extensive informative meetings and consultations enable the population of the region to have their say in planning. This is an essential precondition for the acceptance of the final plan. After the National Committee for Land Management has agreed the plan, management agreements can be concluded. A management plan consists of the following parts:

(a) A description of the wildlife and landscape values that caused designation of the area as environmentally sensitive. The environmental management objectives are established on the basis of this.

(b) A description of the measures with respect to their farm plan and farm management that farmers are expected to take on the basis of these objectives.

(c) The designation of a so-called reference area, i.e. an area in the same region where farm plans and farm management are comparable with those in the environmentally-sensitive area. A reference area is used as a basis for calculating compensation and to ensure that compensation keeps pace with the income potential of farmers who do not conclude management agreements.

(d) A description of the farm plan, farm management and farming conditions in the environmentally-sensitive area as well as the reference area.

(e) Establishing the basis for the levels of compensation that are needed to mitigate the effects of management agreements on farm businesses.

Of course, management objectives must be formulated before management plans can be drafted. The management regulations should be formulated on the basis of advice from the nature conservation bodies, and on the basis of farm management in the area. The objectives, which may vary from area to area, may refer to:

—preservation and development of meadow bird populations:
—botanical management: (a) of grassland vegetation; (b) of vegetation of field margins (e.g. edges of ditches); (c) vegetation of arable lands.
—management of green open spaces between the areas in question and adjacent woodlands and nature areas:
—management for wintering birds (in particular, geese);
—maintenance work on landscape elements, such as wooded banks, pools.

On the basis of these objectives, management provisions are formulated, which are combined into coherent packages of management measures. The farmers can choose one or more of these packages for their farm. Management provisions must meet the following requirements;

—they must make an effective contribution to the preservation and/or development of landscape values;
—their incorporation into the prevailing farm management must not only be possible but technically, economically and organisationally feasible;

—there must be a reasonable balance between the cost of management (management compensation) and its effect on environmental values: a provision with a very high compensation and a relatively minor effect must be prevented;

—compliance with the provisions should be fairly easy to check in practice.

Examples of management provisions are:

—no grazing or mowing of grassland before 15 June;

—no ploughing up of grassland;

—no harrowing and rolling before 15 June;

—no slurry spreading;

—no fertilizer applications on a 3 metre wide margin along field margins

The Memorandum sees wildlife and landscape management as alternative farm produce. A farmer may opt for milk and meat production, but also for wildlife and landscape protection. The problem, however, is the evaluation of wildlife and landscape. The principle is that a farmer who concludes a management agreement should not suffer a decline in income in comparison with a farmer in a reference areas who operates under similar conditions but who produces only food and raw materials. For determining compensation on this basis, descriptions of farm management in the management area and a reference area are needed. Management compensation depends on the extent to which the desired management differs from farm management in the reference area. Farm management for environmental objectives may be associated with:

—a fall in production per ha;

—a higher labour requirement;

—higher costs.

These three elements are the basis for calculating the compensation for each management package. There are four types of compensation:

—*management compensation* for the landuser;

  —*a landowner's compensation* for the owner who does not use the land himself;

—*compensation for losses* in case of a decline in production and over-capacity on a farm;

—*compensation for termination of the agreement* if an area is removed from the list of management areas. The shift from integrated farming to commercial farming may take time. Compensation of this kind has never been paid yet.

Concluding management agreements

After a management plan has been fixed upon, the farmers in a management or reserve area may conclude a management agreement with the Bureau for Land Management. Each quarter, new agreements can be concluded, and each agreement expires six years after the management plan has been fixed upon. At that time, each farmer can decide whether he wishes to renew the agreement for a further six years or to terminate it. If the Provincial Committee sees fit, the management plan can be re-adjusted after six years, e.g. to increase its effectiveness, to improve integration of management or to incorporate new ideas, experience and research results.

Irrespective of this, the farmer can always give notice of termination of the agreement after one year. The farmers appreciate this very much, and the 'trial year' is an incentive to make an agreement. In practice, however, very few farmers give notice of termination after one year.

One problem of the implementation of the regulation is the requirement that the landowner must co-sign the agreement if the land is used by a tenant. In some cases, the former is unwilling to co-operate, because he fears a loss in the value of his land. This fear is unjustified, however, because he receives compensation as well, and the Bureau for Land Management guarantees in the management agreement that it will purchase the land at a price that has not been affected by the integrated management, if the owner appears to be unable to sell it. If an owner sells land that is covered by a management agreement to a third party, the buyer is obliged to take over the management agreement.

Results

Current policy on management agreements did not become operative until 1981, although acquisition of land for reserves had started earlier. Some figures for the last six years are shown in Table 7.2 (numbers of management plans, management agreements and percentage participation).

Particularly in the early years, management agreements were quite a new phenomenon. Recently, farmers have displayed a sharply growing interest in the possibility of concluding a management agreement. The two main reasons for this are:

(i) Though at first there had been much resistance and it was regarded mostly as an attractive option for 'second rate' farmers, management agreements have now become fully accepted. In general, farmers' attitude to them is matter-of-fact or even enthusiastic.

(ii) The super-levy compels dairy farmers to reduce their milk production. This trend makes integrated management easier and more attractive, because the farmer can compensate for the decline in income by protecting wildlife and landscape elements instead.

The first evaluation of management plans after six years has revealed that the size of the meadow bird population has at least stabilized. The average management compensation amounts to some Hfl. 750 a hectare. Moreover, the average size of farms that have concluded a management agreement appears to be equivalent to that of farms without such an agreement. The participants' average age is the same as well, while the intensity of farm management (number of livestock per hectare) on farms with a management agreement is a little lower.

Scope in relation to EEC policy

Some minor adjustments will make the management regulations suitable as national implementation of Article 19 of EEC Regulation 797/85, amendment on 'national support in ecologically-vulnerable areas'. These adjustments will consist mainly of a simplification of the procedures, and co-ordination with Article 15 of the afore-mentioned Regulation on specific measures for hill farmers and farmers in regions with specific problems. This means that a small proportion of the management compensation can be financed from Community funds.

In the next few years, a sharp increase in the area under management agreements is expected.

8 | Portugal

*A.M. Gaspar, E.M. de Sequeira and
A.M. Dordio, Estação Agronomica Nacional
(EAN), Portugal*

Agro-ecological aspects of Portugal

Only 28 per cent of the total Portuguese area is suitable for agriculture, owing to factors such as: shallowness of soil (< 35 cm deep), steep slope, and low fertility level. However, about 55 per cent of the country is farmed and gives a very low crop productivity; for example, about 1,000 Kg of wheat per hectare per year.

The quality of Portuguese soil is impoverished as the altitude increases. North of the Tejo river, the slopes become steeper as the altitude increases, and around 60 per cent of the area is higher than 500 m. In the southern regions, the altitude is generally about 200 m, but is associated with an undulating topography, and much of the soils are on steep slopes of shallow depth and great stoniness.

Soils are therefore of low fertility and mainly acid, of low capacity to store water, difficult to cultivate, and highly susceptible to erosion. This erosion is increased by a long dry season, which encourages areas bare of vegetation. The excess of winter water in shallow soils leads to bad drainage and occasional root suffocation.

The degree of development of the soil improves with decreasing aridity and slope. Bearing in mind the dominant geological characteristics, the divisions are easy: the old massif, the sedimentary Tejo–Sado, and the two ceno-mesozoic fringes. Thus, the distribution of the lithologic foundation of Portuguese soils, in percentage of the total area, is as follows: schistic rocks (33 per cent), granitic rocks (26 per cent), sands and sandstones (15 per cent), limestones and similars (10 per cent) and remainder (16 per cent), which includes alluvial soils and clays from gabbros.

Except in high altitudes, where snow occurs in winter, the climate in Portugal is of the Cs Köppen type, that is, temperate with a hot dry summer generally called Mediterranean climate. Such climate presents two well defined seasons (summer and winter) and two intermediate ones (spring and autumn). The country is well differentiated in terms of climate; in the north, zones with more than 2,000 mm rain, in the south, zones with less than 500 mm; in certain areas of pronounced relief, a microclimate mosaic can also be found.

In Portugal the flora is composed of a mixture of Atlantic deciduous species and Mediterranean and African evergreens. In the northern half of the country there is a predominance of two species of pine (*Pinus pinaster*, maritime pine; and *P. pinea*, European stone pine), and three species of oak (*Quercus robur*, English oak; *Q. toza*, Pyrenean oak; and *Q. Lusitanica*, Portuguese oak). The maritime pine predominates on the sandy coastal soils, and the chestnut (*Castanea sativa*) mainly on the north-eastern mountains; linden (*Ulmus sp.*), poplar (*Populus sp.*), and olive (*Olea europaea*) trees are widespread, mainly the latter.

South of the Tejo river the European stone pine appears in zones of western regions of Ribatejo and Alentego, but the dominant trees are the cork (*Q. suber*) and holm (*Q. ilex*) oaks, and the olive. In the far southern region of the country (Algarve) carobs, almonds, and figs are cultivated. However, figs are also cultivated in the centre near Torres Novas, and almonds in the upper valley of the Douro.

Recently there has been great pressure to cultivate eucalyptus, mainly for paper paste. This increase in the eucalyptus crop has some advantages, namely in terms of exports, but has also many disadvantages, mainly because it substitutes for the natural trees and pasture, leading to human emigration, landscape desertification and destruction of wildlife.

The wild fauna includes indigenous species and introduced species from Central Europe or North Africa. The bigger wild animals found in the mountains are the wolf, wild pig, deer and lynx. The fox, rabbit and hare appear in many zones. Portugal is rich in birdlife, and lies on the winter migration routes of western and central European species. In Portugal 8 per cent of the natural habitat is already protected, but that percentage will probably increase.

Portuguese agriculture is changing very slowly, chiefly because; (a) there is disharmony between the agro-climatic requirements of some crops and the characteristics of the zones where they are produced; (b) low degree of cultural intensification; (c) little use of good quality seeds, fertilisers, and pesticides; and (d) deficient or even improper technologies concerned with soil preparation, sowing, harvesting and storage.

Small farms predominate in the north and central regions, where they are very small indeed (frequently less than one hectare) and composed of many dispersed holdings. In Alentejo, large farms are in an area of drier climate and difficult cultivation.

About one-third of the country is wooded, but the area capable of afforestation is significantly greater. Most of the mountain zones are suited for forestry, grassland or woodland grazing.

Management of grasslands

In the 1930s the 'wheat campaign' was established in order to increase the domestic production of that cereal. That campaign had disastrous effects, chiefly in the southern regions, where erosion and desertification increased significantly and wildlife protection, countryside conservation and the landscape were very much affected.

After the wheat campaign, the soil remained very shallow and its fertility much reduced. Even today the soils in many regions subjected to that campaign are covered by bushes without interest — rock rose, furze, heath. In spite of that, some farmers mainly in southern regions, are ploughing very shallow soils on steep slopes, usually to cultivate cereals alternating with two or three years of fallow, but the yields are insignificant.

South of the river Tejo and in part of the inland centre groves of cork oak and holm oak predominate. Under the trees, which are relatively widely spaced, grows a mixture of native grasses, legumes and other herbs. This pasture and the acorns produced by cork and holm oaks were traditionally used in the autumn as feed for the indigenous pig breed (*porco de montanheira* or *porco alentejano*). But with the appearance of African swine fever and the lack of an efficient treatment of that disease, the *porco de montanheira* has almost disappeared. However, owing to the demand for the excellent traditionally smoked ham and sausages obtained from the *porco alentejano* nourished on oak-groves, and improved knowledge of African swine fever, there are today some initiatives towards the recovery of pig production in this area.

In spite of that, the pastures existing on oak and olive groves (second in importance) are used for sheep, cattle and goats. The natural pastures are of poor quality, as they can support only 0.2 to 2.0 sheep per hectare per year without liming and fertiliser application. Even with liming and fertiliser application, the production increase is very limited in most areas, and this is imputed to the poor quality of the natural vegetation. Indeed this vegetation does not react positively to liming and fertiliser application, perhaps because the local species do not include legumes and grasses suitable for animal feeding. The same happens with the pasture in long fallows before a cereal crop.

With a sown pasture, composed of true legumes (subterraneum clovers, annual medics, yellow lupinus and serradela) and suitable grasses (cocks-foot and Italian ryegrass, for example),and correct liming and fertiliser application, it is possible with recovery of soil fertility to support up to 6 or more sheep per hectare per year.

In the north and western central regions, where the climate is more humid and the soils deeper, there is a preponderance of forests, grasslands and woodland grazings. In the forests, bushes (heaths, furzes, brooms) are frequently found and even some pasture, but these are usually small in area. As in the south, woodland grazing is very important in the north and central regions, since there are frequent associations of trees (oaks, chestnut, elm) with pastures. These, usually of better quality than in the south, are used by sheep, goats and cattle. Livestock raising is most significant in north-western zones, where dairy cattle predominate, but in the north-eastern and central mountainous regions sheep and goats predominate, mainly for milk production.

Proper management of forests and woodland grazing demands comprehensive policies to balance countryside conservation, leisure, recreation and wildlife protection with commercial usage.

Management of wetlands

In the case of wetlands (excluding areas of permanent open water), it is possible to distinguish between areas where there is a permanently high water-table, where agriculture is not possible without proper drainage, and other areas liable to seasonal water-table fluctuations with or without flooding.

The soils located near the estuaries of the main rivers (Vouga, Mondego, Tejo, Sado) and in the southern coastal zone of Algarve are subjected to risks of halomorphism. However, soils located in the valleys of the main rivers but distant from the sea, and those located on their tributary banks are non-saline. Frequently drainage is possible in these kinds of soils, and at least during the summer (and in some cases during the winter too) they can be utilized for the best adapted crops. With the construction of dams and drainage of the soils on the river banks, it is possible to change the crop system based on rice, for example, for others which are better adapted and more productive.

Some wetland zones located near the sea do not have an agricultural capability and are maintained as wildlife reserves (Ria de Aveiro, Estuáurio do Tejo, Ria Formosa). They give protection and refuge to many species including ducks, herons, woodcock and snipe.

Some other wetland zones are economically exploited in a way that does not take into account environmental risks. Thus, wetland zones which are non-saline or bear only a small risk of halomorphism are generally under rice cultivation. The associated use of herbicides and insecticides has contributed seriously to environmental pollution, affecting some habitats and causing a serious reduction of some wild species. Besides attempts to create a few wildlife reserves, wetland management for the conservation of ecological interests has received little care and study.

Today some research is under way in the Baixo Mondego region, on exchanging rice cultivation (in zones where the drainage is easy) for other annual and better adapted crops, like maize (grain or silage), wheat, temporary pasture, forage (for hay and silage), and vegetables.

In the north and central regions, where the winter is colder, livestock, especially sheep and goats, graze the pastures on the mountains during the summer. But in winter they come down to the wetlands (*lameiros* or *prados de lima*), over which water is allowed to flow to avoid freezing.

Management of field margins

In the north-western region (Minho) rains are relatively abundant, the air humidity is high and temperature variations are lower. The region is intensively wooded and is much populated, but the average size of farms is very small (less than one hectare). Irrigated crops predominate, mainly maize, potatoes, and many vegetables, with the non-irrigated crops (rye, forage, pasture) in a secondary role. Vineyards, which are managed on the trellis system and produce the famous *vinho verde*, appear frequently on the boundaries between farms belonging to different peasants.

Usually the farmers have mixed crops, namely annual crops, some fruit trees, vineyards, forest trees, blackberries and flowers. Under the forest trees, bushes and pastures occur. This diversified countryside is very beautiful and provides beauty and recreation, food, pasture, timber and refuge for wildlife. Traditionally the farmers use the branches and needles of maritime pines as source of fuel for domestic use. The bushes were utilized for livestock beds and subsequently for soil manuring. However, owing to the increasing costs of hand labour, today a great percentage of brushwood remains untouched, which is reflected in less organic matter for field and horticultural crops, but wildlife is better protected.

The north-eastern region (Trás-os-Montes) is separated from the Minho by mountain chains, which prevent the humid winds coming from the sea reaching the inland region. The summer period and the rains are shorter than in Minho, but temperature variations are much higher.

There are two distinct zones — one located in the north and near the mountains, called *terra fria* (cold land), and another, located in the south of that zone and comprised of the valleys of Douro river and its tributaries, called *terra quente* (hot land). On the first, the farmers grow mixed crops, mainly potatoes, rye, vineyards, vegetables and livestock. On the second, predominate the vineyards (the best produce the famous port wine, and the others some types of table wines), almonds, olives and some figs.

In the north-eastern part of this region small farms prevail, usually including patches with trees and bushes. These patches act as reservoirs for wildlife in agricultural areas, shield the fields against the winds, provide beautiful places for leisure, give timber, wood, some fruits (mainly chestnuts, hazelnuts, walnuts and blackberries),and gorse for livestock bedding.

The central western region has some resemblance to the north-western region, mainly in respect of demographic pressures and farm size, but forestry is more intensive. It is the main region for maritime pine, but unfortunately those forests suffer greatly from fires during the summer. Olive, cypress, chestnuts, eucalyptus and different species of *Quercus* have an important role,as well as many fruit species. Forage and pasture appear on almost all the farms, especially for dairy cattle, with sheep and goats having a secondary role.

The central eastern region is less populated and frequently the soils are very poor. Forestry plays an insignificant role, although being very common for woodland grazing, mainly for sheep and goats maintained for cheese production. However, the improved meadows and pastures are used especially for cattle feeding. The field margins have an important role. The region being very mountainous, field margins protect the soils against erosion and winds, providing stock-proof barriers and shelter for wildlife.

Quite different is the Alentejo, where plains predominate, with little rain and a hot, dry and long summer. Trees are scarce because all the region is dominated by the wheat crop, with barley, oats, chick-peas and sunflowers as secondary crops. The dominant trees, which sometimes break the monotony of the landscape are cork and holm oaks and olive, but in recent years interest in the eucalyptus has increased, as in the centre of the country. Stunted bushes (mainly rock rose) grow predominantly in the shallow soils, mainly in the eastern zones (near Mértola), one of those most affected by the wheat campaign in the 1930s.

The countryside is arid, the population restricted and the limited wildlife almost confined to the mountains, to the zones of shallow and infertile soils during the fallow periods, or to the occasional field margins located near the rivers and streams.

The northern zone of Algarve has an austere landscape because it is a mountainous region, where the dominant trees are the cork and holm oaks, and eucalyptus. The rock rose is very frequent and the wildlife is diversified. Farmers sometimes grow oats or barley with poor results following long fallow periods. The pastures are of poor quality and usually for goats and sheep. This zone suffered very much from the wheat campaign in the 1930s.

Quite different from the north zones of Algarve, are the most populated zones in the south of the same region, known throughout the world for their tourist interest. The farms are again very small and diversified. Whenever the farmers have water for irrigation, usually from wells, they grow protected crops (tomato, melon, pepper, cucumber, bean, egg-plant and flowers), citrus fruits and vegetables. When irrigation is impossible, the farmers use mixed cultivation — almonds, carob, olive, wild pistachio, oak and fig trees, winter cereals and pastures for livestock.

The field margins may exercise an important role against serious pests and diseases, but they also can have adverse affects, namely in spreading potential pests, weeds and diseases into the crops. The situation varies, depending on different biotic and ecological combinations. The diversity of the flora implies a great variety of the corresponding entomological fauna, and this must be studied in order to help natural control of pests and diseases of cultivated plants. Each case must be carefully studied, with detailed observations in the field. As examples, we can say that in Portugal:

(a) Hop (*Humulus lupulus*) is attacked by the aphid *Phonodon humuli*, which survives during the winter on certain plants in field margins. By eliminating these plants, aphid attacks on the hop crop can be considerably reduced.
(b) In order to obtain control of the walnut tree (*Juglans regia*) aphid (*Chromaphis juglandicola*), it is recommended to have some oak species near the walnut orchards. On the oaks are developed not only other aphid species, which do not attack the walnut trees, but also the natural parasites of the walnut tree aphid. This is reflected in a substantial decrease in the aphid population on the walnut trees.
(c) *Pyracantha* (*Pyracantha sp.*) is the alternate host of an aphid (*Aphis pomi*), which attacks apple trees. Controlling pyracantha significantly reduces attacks by the aphids on apple trees.
(d) On the oleander (*Nerium oleander*), attacked by *Aphis nerii*, a natural complex of predators and insect parasites occur which attack *Aphis citricola*. Oleander presence near citrus orchards significantly contributes to the control of *A. citricola*. However, the presence of oleander near citrus and olive tree orchards may be disadvantageous in other ways, because it harbours *Saissetia oleae*, a serious pest of citrus and olive.

(e) A section of the tomato industry provides tomato seedlings to farmers. In the past it has been found that in nurseries located near private kitchen-gardens the presence of leaf hoppers and of viruses transmitted by them has led to infection of the tomato seedlings. Moving the nurseries to areas separated from the kitchen-gardens by field margins of a sufficient size has enabled the tomato seedlings to be grown free from virus and leaf hoppers in the nurseries.

References

Gaspar, A.M.; Henriques, F.S.; Moreira, T.J.S.; Serafim, F.D. April 1987. *Utilization of the Portuguese Soils for Non-Food Crops (Excluding Wood Production)*. Report requested by JNICT under the FAST II Programme of the European Community. Lisbon.

Ilharco, F.A., 1983. A Seccäo de Equilibrio Biológico de Afidios do Departamento de Entomologia da Estac[um]ao Agronómica Nacional: Objectivos e Realizacóes. *Bol. Soc. port. Ent.* 32 (II-2).

Sequeira, E.M., 1987. *Da Necessidade do Aumento da Area das Pastagens. Pastagens e Forragens* (in press).

West Germany

E. Lübbe, Der Bundesminister für Ernahrung, Landwirtschaft und Forsten, Bonn

Management of grassland

Thirty-eight per cent (4.6 m ha) of the total agricultural land in the FRG is used as permanent grassland. The mean yield of natural grassland is 80–85 dt/hay equivalent and supplies about half of the livestock need. The amount and spatial distribution of grassland in West Germany varies from region to region and in response to environmental factors. Precipitation and range of temperature affect grassland as well as soil conditions, especially:

— in the coastal area of Northern Germany;
— in rainy hilly areas and in the Pre-Alpine zones;
— in river valleys with high groundwater table (see Figure 9.1) In the southern parts of Germany meadows are predominant, in the northern ones we find more pastures.

Natural environmental factors together with traditional agricultural practices have led to typical plant societies of permanent grasslands characterised 'molinietalia' (wet meadows) and 'arrhenatheritalia' (neutral meadows and pastures). Extensive forms of grassland such as scattered meadows, alpine pastures and rough pastures are now of little agricultural importance, and amount to only 3 per cent of permanent grassland. This seems too small an area for the needs of nature conservation and environmental interests.

Der Bundesminister fur Ernahung, Landwirtschaft und Forsten, Postfach 140270, 5300 Bonn 1, West Germany.

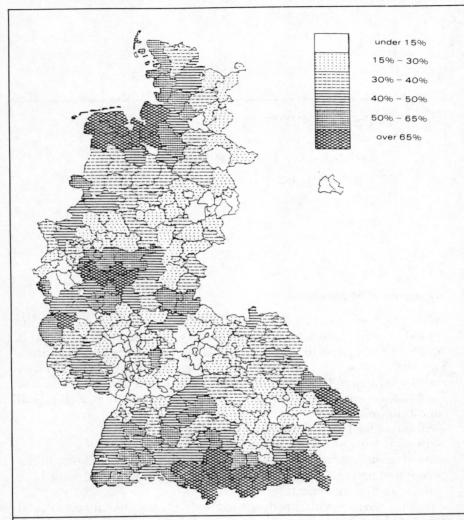

Map 1.   Regional amount of permanent grassland
as % of total agricultural land.

Figure 9.1 Permanent grassland as percentage of total agricultural land
in West Germany by regions

Future development

Recent advances in harvesting and conservation of grass are described by Zimmer (1985). Under pressure of surplus production and the agricultural political situation which especially affects milk production enterprises, the following changes may occur:

— It is estimated that up to 1 m ha of permanent grassland will be redundant.
— Marginal grassland locations will be given up first. It is proposed to redevelop these areas into natural biotopes, necessarily combined with some permanent landscaping.
— Extensive forms of animal breeding will improve animal production.
— At the same time biological and technical developments will enable farmers to use productive grasslands more intensively.

Management of wetland

Wetlands should only be utilized as extensive grassland as already described. If drained arable land could be changed back into grassland especially near open water courses and lakes, there would be many advantages:

(1) 'Biotope networks' could be established.
(2) Erosion problems would be minimized with the consequence that eutrophication problems in surface waters would be reduced (less nutrient inputs).
(3) Natural conditions in small rivers would be improved and self-cleaning possibilities increased.
(4) Flood protection methods could be changed with fewer regulating reservoirs in the upper parts of water courses, and more controlled flooding of areas downstream.

However, these proposals would have to be paid for, and the public would have to buy or rent these areas in order to manage them. In Germany this could be done by associations for water resources and land improvement.

Bogs and fens should also be put into an integrated nature conservation scheme. Conservation and regeneration of peatlands are possible where the climatic situation is suitable, appropriate hydrological conditions are met and pollution can be avoided. The residual parts of moors can only be saved if special protection zones are established. More details will be found in 'The capacity to Protect Bogs in Northwest Germany' by Kuntze and Eggelsmann (1981).

Management of field margins

Agricultural ecosystems are often patchy environments, and include habitats of different kinds; some are unproductive and unmanaged. Nevertheless, they fulfil different functions in the countryside and their importance increases with the intensification of the management of productive land. The following small biotopes can be included under the topic of 'field margin':

(1) 'linear' structures, such as
  — hedges
  — field balks and waysides
  — unpaved agricultural roads
  — water courses
  — banks
  — tree rows
  — dry ditches.
(2) 'insular' structures, such as
  — ponds
  — wetlands
  — bogs, fens
  — field copses
  — tree groupings
  — fallow land
  — single trees.

These elements are an integral part of the countryside and the surrounding agricultural land, but there is a lack of information concerning their nutrient requirements, microclimate conditions, and the distribution and diversity of the flora and fauna included in them, which has not yet been fully classified. Three examples illustrate some of the factors relevant to such biotopes:

(1) changes of wind velocity, transpiration, precipitation and dew forming behind hedges (Figure 9.2);
(2) Functions and relations among hedges (Figure 9.3);
(3) Activities of ground beetles in winter cereals (Knauer and Stachow, 1986).

As a preliminary result of research, it can be stated that the above mentioned linear structures should form a network along the edges of fields, which should be of a maximum size of 300 × 600 m if hedges are at least 3 m wide (Figure 9.4) and balks have a minimum size of 1 m (Knauer and Stachow, 1986;Knauer, 1986). The optimum distance to other ecological features like field copses should not exceed 400 m.

In West Germany many states (*Länder*) have special programmes

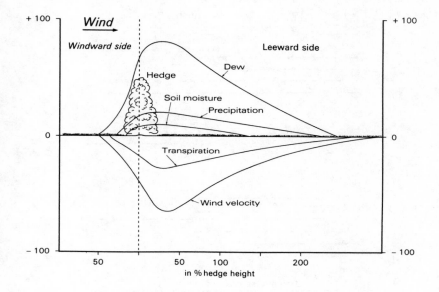

Figure 9.2 Changes of wind and velocity, transpiration, soil moisture, precipation and dew related to the distance behind an isolated hedge (in percentage of hedge height)

concerning field margins. The different programmes and their costs are listed in Table 9.1. It seems that more and more farmers are accepting the possibility of earning non-agricultural income in these ways, but any positive results depend on sufficient funding.

A general overview of the problem discussed is given in the report entitled 'Land Use and Nature Protection in the Federal Republic of Germany' by Conrad (1986).

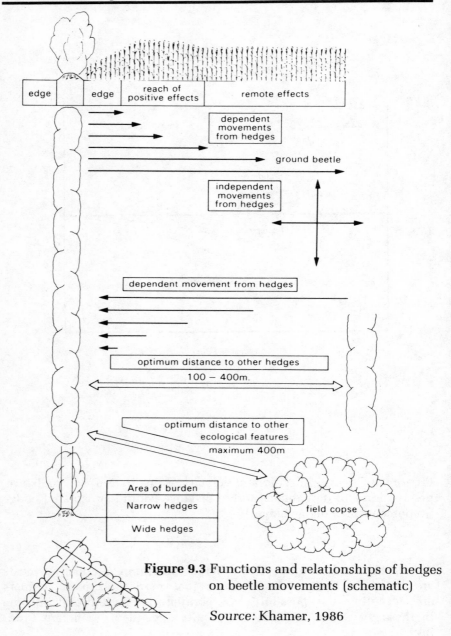

Figure 9.3 Functions and relationships of hedges on beetle movements (schematic)

Source: Khamer, 1986

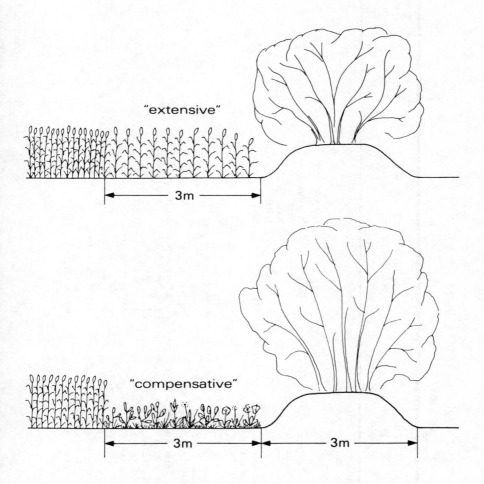

"extensive"

3m

"compensative"

3m   3m

Figure 9.4   Ecological improvement of hedges combined with extensive use of edges (above) or by planting weeds (below)

Source: Knauer and Stachow, 1986.

Table 9.1 Programmes of selected West German states with direct or indirect implications for nature conservation

State	Programme	Funds available	Areas affected	De-intensification of agriculture	Concept for biotope protection
Bavaria	1. Bavarian programme for alpine and low mountain regions, parts A,B,C	DM 10m (1985)	Potentially large, but financial restrictions	In certain areas	No
	2. Liquid manure programme	DM 16m (1984)	Potentially large, but financial restrictions	—	No
	3. Campaign: 'More green through farmland consolidation	DM 3.5m (1985)	Small	—	No
	4. Handicap compensation payment according to Art. 36a Bavarian Nature Conservation Act	DM 1.17m (1985)	Small	Yes	No
	5. Programme for the protection of meadow-breeding birds	DM 2.6m (1983—4) DM 1.5m (1985); DM 1.5m (1986)	Small	Yes	No
	6. Landscape management programme	DM 1.75m (1985); DM 2.5m (planned for 1986)	Small	No	Yes
	7. Programme for the promotion of fertiliser and herbicide-free buffer zones around fields and pastures	DM 35,000 (1985); DM 390,000 (1986); more than DM 1m planned for 1987	Small	YesNo	

Table 9.1 continued

State	Programme	Funds available	Areas affected	De-intensification of agriculture	Concept for biotope protection
Hesse	1. (a) Programme for the promotion of pesticide-free buffer zones around fields	DM 800,000 planned for 1986 for 1a and 1b	Potentially large	Yes Yes	Long-term aim of both programmes is the establishment of a network system of biotopes
	(b) Programme on 'eco-meadows'		Restricted to certain kinds of of meadows	Yes	
	2. Guidelines for granting compensation to farmers giving up milk production and for for the allotment of the extra milk quotas at disposal	DM 9.9m (1986) depending	Potentially large on funds available	Yes	
Lower Saxony	1. Guidelines: handicap compensation grants	DM 1.85m (1985) DM 3.4m 1986	Very small, Very small, 6,200 ha (1985)	Yes Yes (in these small areas)	No No
	2. Programme similar to programme 1(a) in Hesse	DM 100,000 (planned for for 1987)	Small	Yes	No
Rhineland-	1. Equalisation payment	DM 18.01m	Potentially	Yes	No

Table 9.1 continued

State	Programme	Funds available	Areas affected	De-intensification of agriculture	Concept for biotope protection
Palatinate	on the basis of the EC directive on farming in less favoured areas	(1985)	large depending on funds available		
	2. Programme for the protection of species and biotopes				
	(a) Programme for the protection of field border structures	DM 25,000 (1986–7)	Potentially large	Yes	Long-term aim of all
	(b) Programmes for the de-intensification of agriculture	DM 25,000 (1986–7)	Potentially (funds!)	three Yes is the	programmes establish-
		Potentially	large (funds!)	ment of a	network system of biotopes
	(c) Linkage of biotopes worthy of protection	DM 2m (1986–7)	Potentially large (funds!)		
	3. Preparation of areas for nature protection on the basis of farm-land consolidation	DM 1.7m (1984) DM 1.25m (1985) DM 4.2m (1986–7)	Very small, 115 ha (1985)	Yes	Yes
Baden-Würtemberg	1. Biotopes and landscape management programme	DM 4.4m (1984) DM 5.2M (1985)	Large	No	Yes

Table 9.1 continued

State	Programme	Funds available	Areas affected	De-intensification of agriculture	Concept for biotope protection
	2. Purchase of areas important for nature protection	DM 2.78m (1984) added subsidies to communities and for this purpose DM 715,000	Very Small	?	?
Schleswig-Holstein	1. Programme for the de-intensification of agriculture	DM 300,000 (1985)	Small	Yes	No
	2. Programme for nature protection and structural improvements on the border between Schleswig-Holstein and Mecklenburg (GDR)	DM 5m (planned 1986)	Potentially large (funds!)	Yes	Yes
North Rhine-Westphalia	1. Programme for the protection of wet meadows (a) Urgent programme for nature protection in 1985/86	DM 66.4m (1985–86)	Restricted, but large	Yes	Yes
	(b) Purchase of areas for nature protection			Yes	Yes
	2. Programme similar to programme 1(a) in Hesse	DM 45,000 (1985–6)	Small	Yes	Yes
	3. Low-range mountain programme	DM 1m	Potentially (planned in 1987) restrictions	Yes large, but financial	(Yes)

Monitoring landscape and wildlife habitats

See the report of Knauer and Stachow (1986) where examples of research results are given.

References

Conrad, I. 1986. *Land Use and Nature Protection in the Federal Republic of Germany*. Wissenschaftszentrum Berlin für Sozialforschung, Internationales Institute für Umwelt und Gesellschaft, II UG dp 86–16.

Knauer, N. 1986. Okologische und landwirtschaftliche Konzepte zur Verwendungfreigesetzter Flächen. *Neues Archiv füur Niedersachsen, 35,* 229–43, Göttingen

Knauer, N.and Stachow, U. 1986. Verteilung und Bedeutung verschiedener Strukturelemente in einer intensive genutzten Agrarlandschaft. *Verhandlungender Gesellscahft fur Ökologie, 14,* 151–6.

Kuntze, H.and Eggelsmann,R. 1981. Zur Schutsfähigkeit nordwestdeutscher Moore. *TELMA, 11,* 197–212, Hanover.

Zimmer, E. 1985. Statusbericht der Grundfutterversorgung in der Bundesrepublik Deutschland. FAO/OECD Genf, unpublished.

10 | Great Britain

*J.M. Way, C.J. Feare and I.M. Tring,
Ministry of Agriculture, Fisheries and
Food, London and Worplesdon, Surrey*

Management of grassland

For the purposes of this chapter all grasslands in Great Britain (England,
Scotland and Wales) except those of montane regions, maritime cliff
edges and subtidal areas of salt marshes have been created directly or
indirectly by man, and (with these exceptions) natural grasslands, in the
sense of grasslands untouched by man, do not exist (Wells, 1983). All
the grasslands with which we are concerned are ecologically unstable,
and in the absence of management will develop, over differing periods
of time, into woodland. Whilst this is a broad generalisation, it is true
for all practical purposes.

Objectives

Grassland in Great Britain is managed primarily for agricultural
production, but some grasslands have historically been managed for
landscape (amenity) and recreation, and more recently in the twentieth
century for the conservation of wildlife. Other significant areas, although
small in comparison with agricultural use, have arisen as a secondary
consequence of other land uses, including the banks and verges of roads,
motorways and railways, and on military training areas.

Agricultural objectives over the last half century have been to increase
the efficiency of agricultural production, involving intensification of use
of grassland and significant conversion of grassland to arable.

Environmental objectives for landscape are difficult to define, but
basically concern the harmony of the countryside and control of
intrusion of disturbing elements. Wildlife conservation is concerned
with protecting and sustaining populations of wild plants and animals

Table 10.1 Areas of agricultural grasslands and rough grazings in Great Britain

	% of total area	% of agricultural area
Great Britain		
Grasslands and rough grazings	55	73
Rotational and permanent grass*	28	37
Rough grazing	27	36
Area (ha m)	23.0	17.6†
England and Wales		
Grasslands and rough grazings	47	63
Rotational and permanent grass*	36	47
Rough grazing	12	16
Area (ha m)	15.2	11.5†

Source: Derived from Brockman (1982)
*Equivalent to improved or improvable productive grasslands.
†77 per cent of the total area.

(NCC. 1984). These and other objectives for amenity, recreation and wildlife conservation have been articulated with mounting concern (and political pressures) following in the wake of agricultural improvements, as it has become increasingly apparent that, rather than agriculture maintaining valued aspects of the countryside, it has caused changes that have been seen to be damaging. Environmental concerns have been spurred on in the last decade as agricultural improvements have led to production surpluses. Consequently, there are conflicts between the objectives of agricultural production and other interests; but there are also conflicts between environmental interests (Brotherton, 1977), notably those of recreation and wildlife conservation.

Occurrence

A broad distinction is made in Great Britain between lowland and upland grasslands. This is primarily based on agricultural criteria, but has some basis in floristics as discussed by Ratcliffe (1977). Grasslands are divided by Brockman (1982) into:

(a) rotational, temporary, grass within an arable rotation, principally in the lowlands; and permanent grass in fields or relatively small enclosures, not in an arable rotation, in the lowlands or the uplands;

(b) rough grazing, uncultivated grasslands, wholly enclosed or in relatively large enclosures found on moorlands, heaths, hills, uplands and (chalk or limestone) downlands.

Figures for these are given in Table 10.1.

These figures give orders of magnitude (and should only be interpreted as such) for agricultural grasslands and rough grazings showing that they occupy about half the surface of Great Britain, and three-quarters of the agricultural land. In England and Wales the figures are lower, largely reflecting the higher proportion of other land uses and of other crops, mainly in England. It is also notable that the proportion of rough grazing in England and Wales (of which most is in Wales) is about half that for Great Britain, reflecting the lowland character of much of England and probably the greater intensity of management of grassland in England and Wales, compared to the large area of rough grazing in Scotland.

The wet oceanic climate and the physiography of the west and north of Great Britain are particularly suited to the growth of grass, whilst Eastern England with a drier climate, few hills and no land over 300 m is predominantly arable — of the ten counties in this region permanent pasture and rough grazing occupy less than 20 per cent of the farmland (Duffey *et al.*, 1974).

Types of grassland

Agricultural. Agricultural grasslands, depending upon their status as rotational, permanent or rough grazing, comprise a continuous range of variation from good quality *Lolium perenne*/*Trifolium repens* swards, through swards of increasing diversity with varying proportions of *Poa trivialis*, *Agrostis spp*, *Dactylis glomerata*, *Festuca ovina*, *Phleum pratense*, *Cynosurus cristatus*, *Deschampsia cespitosa* and *Alopecurus pratensis*, to mainly *Agrostis*, and in rough grazings from *Agrostis* with *Festuca rubra* to poorest quality grasslands of *Calluna* with *Molinia caerulea* and *Nardus stricta* (Brockman, 1982). These grasslands may be classed as meadows, cut for grass conservation as hay or silage; as pastures for grazing of stock; or as combinations of these uses. The farmer is concerned not only with the growth of his grassland but with its utilisation by his stock, measured as milk production, weight gains, or by condition (e.g. sheep at ovulation and at lambing). He will use his skill to achieve his objectives through management of soil, the sward and his animals. Essentially he will be looking to improve soil fertility by drainage and use of fertilisers, and of lime to correct acidity, and the sward by manipulating the timing and intensity of defoliation and by the selective use of herbicides. The results of good agricultural management will be a strongly growing grass sward comprising a small number of preferred species and few broadleaved flowering plants.

Environmental. Ecologically grasslands are categorised as acid, basic or neutral, qualified hydrologically as wet or dry, by physiographic features, and floristically. The distant landscape interest of grasslands is mostly satisfied by the grassland *per se*, but close-up the detailed components are important, and colourful herb-rich swards are a great pleasure to look at. The landscape interest is therefore determined by the perspective of the viewer. In terms of wildlife conservation, where populations of species and the intra- and inter-relationships of varied communities of plants are of fundamental interest, most forms of modern agricultural management of grasslands are damaging. Figures quoted by the Nature Conservancy Council (1984) show that of lowland neutral grasslands (including herb-rich hay meadows), 95 per cent are now lacking significant wildlife interest, and only 3 per cent are left undamaged by agricultural intensification. For lowland grasslands of sheep walks on chalk and limestone 80 per cent have been lost or damaged, mainly since 1940.

Remedies

In Great Britain agriculture will continue to be the major land use of grassland. However, there are clear indications of changes in the distribution and intensity of agricultural use, so that there are opportunities to introduce management protocols that are sympathetic to environmental interests. This recognises that traditional (pre-1940) agricultural practices were largely sympathetic to landscape and nature conservation, and as a corollary that forms of agricultural management are still required for grassland for environmental objectives. However, this implies a change from high-input to low-input systems, which in turn requires economic assistance to the farmer in order to maintain his income. Provisions for voluntary management agreements with conservation payments exist in National Parks, in Sites of Special Scientific Interest administered by the Nature Conservancy Council, and from Local Authorities. Purely voluntary arrangements are in the process of coming into operation in Environmentally Sensitive Areas in Great Britain, in which farmers are to be paid a fixed sum per hectare by the Agriculture Departments to undertake certain practices, including in the case of grasslands, not to use inorganic fertilisers, to restrict use of organic fertilisers, not to use herbicides except under specified conditions, not to cut for hay or make silage before certain dates, and to control the period and density of stocking of animals. These constraints are based on experience and the best information available, but there are many uncertainties, and there are requirements for research on the agricultural management of grassland for environmental benefits.

Field margins

The intensification of arable agriculture in Great Britain, especially since the early 1940s, led to a considerable reduction in the country's hedgerows. Hedge removal gained impetus with the need for more land for food production and with the development of larger farm machinery that was more economical for use in large fields. While hedges are a man-made artefact of the British countryside, they are nevertheless regarded by the public as 'traditional'. In fact, hedgerow loss transforms the landscape so much that aesthetic considerations have probably played a larger part than ecological concerns in public attitudes towards hedges and their contribution to the countryside. However, work by the (then) Nature Conservancy described many aspects of the biological interest of hedges in the 1960s and 1970s, and further work by many organisations has confirmed this subsequently. More recently, food surpluses have taken the pressure off land use, and conservationists and land-use planners have begun to re-examine the role of hedgerows in the wider context of striking a balance between agricultural production and environmental concern over field margins as wildlife reservoirs. In addition, declines in some aspects of wildlife diversity on arable land have stimulated specific areas of work aimed at minimising deleterious effects of intensive agriculture around field boundaries.

Studies on a variety of aspects of field margin ecology and agronomy are currently being undertaken in Great Britain and, while there has been little co-ordination between these studies in the past, the Agricultural Development and Advisory Service (ADAS) of the Ministry of Agriculture, Fisheries and Food is now attempting to direct research on field margins towards a hierarchy of priorities discussed at meetings with interested organizations. Field margins do, of course, serve a number of agricultural functions and the broad aims of current work are to reconcile agricultural uses with the maintenance of the greatest wildlife diversity possible.

Following the realization of the environmental problems caused by organo-chlorine pesticides, pesticide application has continued to cause concern and two studies in particular are investigating the effects of reduced pesticide input. The ADAS Boxworth Project (at Boxworth Experimental Husbandry Farm, Cambridge) investigates the effects of three levels of pesticide application on crop yield and on a broad spectrum of wild animals and plants, ranging from arable weeds and insect pests to birds and mammals. While the pesticide application regimes are applied to the whole crop, the environmental responses within field margins are being monitored by specialists from a number of UK institutions.

The environmental benefits of reduced pesticide spray input, in terms of leaving a 6 m strip around the field edge unsprayed, are being assesssed in the Game Conservancy Cereals and Gamebirds Research Project. This study had its origins in research on reasons for the decline of the Grey partridge (*Perdix perdix*), where it soon became apparent that the decline centred on a scarcity of insect food for chicks brought about by herbicide and insecticide use. Partridges have also been found to require a dense herb flora within the hedge bottom for successful breeding and an uncropped margin is beneficial to chick survival. The leaving of a 6 m unsprayed strip around a cereal field edge allows the development of a diverse herb flora within the crop. This has its own intrinsic interest and the project has further diversified to incorporate the responses of butterfly populations to the unsprayed margins.

The effects of uncropped strips on agronomic and wildlife parameters have been assessed in several ADAS studies concentrated on Experimental Husbandry Farms. These studies are now being extended and co-ordinated with a single project on six farms, initially investigating the responses of crops, weed flora and insect fauna in the hedge and crop edge to the presence of marginal uncropped strips maintained respectively by mowing, rotovation and herbicides. The plots used in the study are at present small but it is planned, when larger plots are used, to include small mammals and birds in the investigation.

Mammal and bird populations are being studied independently by ADAS and the British Trust for Ornithology (BTO). The former set out to quantify small mammal cycles in hedgerows, but the aims have now been redefined to assess the impact of hedge plant species diversity on small mammal abundance and movement in different parts of Britain. The BTO, through its Common Birds Census scheme, is investigating the effect on breeding bird populations of many aspects of hedge structure. These studies are being further extended to investigate the importance of hedges to birds in winter.

Hedgerows are used as overwintering sites by many insect species and the importance of these winter refuges to the development of spring and summer populations is of particular relevance to the possibilities for biological control. Polyphagous insects move out from the hedge into the crop with potential benefit in the control of cereal insect pests, and the dynamics of this system are being investigated at Southampton University.

Many other institutions, including universities, voluntary bodies and publicly funded research organisations are undertaking research or demonstrations of the management of field margins for a variety of purposes. However, the principal aim of management will vary from farm to farm according to local interests, and the collaborative approach adopted in many studies will help to identify the most beneficial compromises for farming and wildlife interests. Ultimately, the objective of research is the production of an advisory package that can be used by farmers to identify management options and their cost appropriate to given circumstances.

Lowland wetlands

The majority of important wetlands in Great Britain, because of the length of coastline, are located on or near the coast (Baldock, 1984). They include river estuaries, salt marsh, coastal grazing areas and on a smaller scale the remnants of bog, fen and marsh (Figure 10.1).

In Table 10.1 it has been shown the agriculture occupies about 76 per cent of the surface of Great Britain. With agriculture occupying such a large proportion of the total area together with the attendant Agricultural Policy, agriculture inevitably exerts a considerable influence on the wetland environment.

Definition of wetlands

The term 'wetlands' is somewhat arbitary and no single definition has been universally accepted. Perhaps the most widely used is that embodied in the Ramsar Convention which defines wetlands as: 'areas of marsh, fen, peatland or water, whether natural or artificial, permanent or temporary with water that is static or flowing, fresh, brackish or salt, including areas of marine water the depth of which at low tide does not exceed 6 metres'. This is a broad definition but does appear to exclude wet or periodically flooded lowland pasture and areas of reclaimed marsh, where the water-table is permanently high, typically within 20 cms of the surface.

This generic use of the word is important, but in every day use 'wetlands' often refer to a much smaller range of habitats including wet meadows and places with saturated and often flooded soil. Precisely

Figure 10.1   Wetlands in Great Britain

Source: Baldock, 1984.

where these habitats begin and end may be difficult to determine (Baldock, 1984).

Conflicts

Wetlands are often vulnerable habitats and may be threatened from many different quarters. Because many plants and animals are dependent on particular hydrological conditions and are sensitive to changes, drainage has a marked effect. Other principle causes of 'damage' are industrial reclamation, pollution from agricultural and urban sources, as well as peat exploitation, agricultural developments and chemical enrichment of lowland sites in partiuclar (Baldock, 1984; Newbold et al., 1983).

Land drainage

The roots of land drainage extend back 2,000 years. Interest has ebbed and flowed and it is no accident that periods of activity have coincided with demands for increased food production. The term 'land drainage' is taken to include both arterial and field drainage. Most arterial schemes designed to benefit agriculture involve a lowering of neighbouring water-tables to reduce seasonal flooding and make better provision for field drainage (Trafford, 1982).

However land drainage aims to change the soil water regime and hence it is not surprising that this is one of the sharpest points of conflict between agriculture and conservation. On any particular wetland site a modest change in drainage status may have serious consequences for conservation. In many cases it may not be possible to reach some mutually acceptable compromise without dividing the area into segments where agriculture or conservation may be given a greater or lesser degree of priority. To give the matter some proper perspective, however, there are in fact only a very few cases per year which cause problems. The reason for this is that the vast majority of the work, whether field drainage or arterial, is concerned with up-grading land which is already in arable farming. Thus in virtually every case an acceptable compromise, or an opportunity to enhance the conservation interest, is possible.

The impact of drainage on landscape is varied. On the one hand tree and hedge removal, ditch elimination and channel straightening, as a part of a drainage operation, can be damaging. On the other hand, with sensitive design and consultations with other environmental interests, the impact of drainage on landscape can be minimised, and on many occasions has enhanced it.

Agricultural policy

Agricultural support under the Common Agricultural Policy has encouraged farmers in Great Britain to drain and improve wetland sites, including ploughing them for cereal production in order to increase food production and profitability (Trafford, 1982). The Government has recognised that many such sites are of significance at either the international or national level because of their wildlife value and in particular as habitats for wild birds. It has therefore introduced measures to protect such sites.

In 1976 the United KIngdom ratified the Convention on Wetlands of International Importance especially as Waterfowl Habitat (the 'Ramsar' Convention), and a programme of designation of relevant sites in the United Kingdom has been undertaken. This programme is continuing. Measures to provide protection are laid down in the Wildlife and Countryside Act 1981, whereby important sites are notified as Sites of Special Scientific Interest and proposed changes in management, including those in agricultural operations, which may adversely effect the wildlife value of such sites must now be notified in advance to the Nature Conservancy Council. The Council is empowered to enter into management agreements with owners or occupiers of the land in order to protect or enhance its wildlife value. If necessary compulsory purchase powers can be used. The cost of protecting important wetland sites thus falls on the public at large rather than on the individuals who depend upon the land in question for their livelihood (Castle *et al.*, 1984).

Implications for research

There is a need to assess the impacts of changing land use, habitat management and drainage control, and populations and distribution of selected plant and animal species.

References

Baldock, D. 1984. Wetland Drainage in Europe. International Institute for Environment and Development *and* Institute for European Environmental Policy.

Brockman, J.S. 1982. Grassland. In *The Agricultural Notebook* (ed. R.J. Halley). Butterworth, London. pp. 173–202.

Brotherton, D.I. 1977. Lowland grasslands. In *Conservation and Agriculture* (eds J. Davidson and R. Lloyd). Wiley-Interscience, Chichester. pp. 81–93.

Castle, D.A., McCunnall, J., Tring, I.M. 1984. *Field Drainage Principles and Practices*. London: Batsford.

Duffey, E.A.G., Morris, M.G., Sheail, J., Ward, L.K., Wells, D.A. and Wells, T.C.E. 1974. *Grassland Ecology and Wildlife Management*. Chapman and Hall, London.

Nature Conservancy Council. 1984. *Nature Conservation in Great Britain*. The Nature Conservancy Council. Peterborough.

Newbold, C., Purseglove, J., Holmes, N. 1983. *Nature Conservation and River Engineering*. Peterborough: Nature Conservancy Council.

Ratcliffe, D.A. 1977. *A Nature Conservation Review*. Cambridge University Press, Cambridge.

Trafford, B.D. 1982. The background to land drainage policies and practices in England and Wales. Seminar L'Ecole Polytechnique, Paris, February 1984.

Wells, T.C.E. 1983. The creation of species rich grasslands. In *Conservation in Perspective* (ed. A.Warren and F.B. Goldsmith). Wiley-Interscience, Chichester. pp. 215–232.

Management of
Field Margins

Introduction

C.J. Feare, Ministry of Agriculture,
Fisheries and Food, Worplesdon, Surrey

Part III concentrates on the environmental aspects of boundaries to arable land. In a discussion on wildlife problems it is easy to overlook the important function of agricultural land, so agronomists and environmentalists should aim to develop integrated objectives, seeking common ground but accepting that some compromise may be necessary.

One theme comes through Part III: that of diversity. Webb shows that hedgerow networks still exist at high density in many parts of Ireland and Baudry highlights the importance of the network system, and especially interconnections with other habitats. Marshall demonstrates the need for high diversity and thus stability in the boundary flora and Wratten shows how floristic diversity on specially created banks could increase invertebrate populations. These latter two chapters, supported by the Portuguese national report, further show how this diversity could be of benefit to agriculture. Clearly, diversity is something that should be sought in agricultural systems and would go a long way towards satisfying the public desire for 'traditional' landscapes, a need highlighted by Tinker. We do, however, have to think of diversity on widely differing scales, from floristic diversity in each square metre of hedgerow to diversity of land use, permitting interactions between different agricultural and wildlife habitats: this was, after all, the basis of rotational systems of agriculture that were used to optimise land use for many decades.

In order to seek wide-ranging benefits from research on environmental diversity, the research clearly needs to be multi-faceted, covering environmental functions, agricultural functions and, of course, cost of operations and profitability of the land; some of this profitability can be readily measured, as in crop yields, while other profits, such as public appreciation and awareness, are less quantifiable. Proposed environmental recommendations should not be inimical to agriculture: a balance needs to be sought.

Multi-faceted research needs co-operation and co-ordination between an array of disciplines. This has been sought in the United Kingdom, where a series of meetings have been held recently, involving representatives of widely differing interests, in order to determine research priorities. In the United Kingdom research on the hedge structure and its management, pesticide spray drift, boundary strips and unsprayed crop edges were some of the areas identified as in urgent need of strengthening. With the emergence of surpluses the last two of these, involving alternative forms of land use that can be beneficial to both agriculture and conservation, are increasingly becoming realistic possibilities.

On a European scale, problems differ in different regions and Part II (the National Reports) indicates variation between Mediterranean and more northern areas. However, this Workshop has clearly demonstrated that important research is taking place in several EC countries but that knowledge of this research seems not to cross national boundaries. While the Workshop has not indicated major areas of duplication, it has highlighted areas of complementary research, for example the fine scale work undertaken in the United Kingdom and the landscape aspects studied in France. There would be much benefit for closer liaison between these studies, co-ordinating the work and integrating the results. Future workshops could have a useful role here.

At least two areas appear to be receiving little study at present. First, the role of field margins, in both agricultural and environmental terms, in predominantly grassland areas seems particularly neglected; co-operation between researchers in the United Kingdom and Ireland should be a distinct possibility here.

Second, we need better understanding of how to convince farmers that ecologically sympathetic management of field boundaries (and cropped areas) will provide better long-term contingency use of chemicals, with consequent unappreciated effects. This is a problem that merits consideration throughout Europe.

Hedgerows and hedgerow networks as wildlife habitat in agricultural landscapes

J. Baudry, Institut National de la Recherche Agronomique, Saint-Pierre-sur-Dives

Land is a limited resource with many uses (agriculture, forestry, recreation, nature conservation, construction); this can be a source of conflict. Some of these uses can be separated in space, while others cannot (e.g. land which is part of a watershed may influence water quality). Planners and policy makers now have to decide whether to use land for a single purpose or whether to maximise its uses through some form of diversification, especially as it is forecast that millions of hectares will be taken out of production in EC countries within a few years. Do we want productive land on one side and pseudo-wilderness on the other, or is there a way to plan, design and manage landscapes for multiple purposes, using sound ecological principles?

Though more research is required, I think that multiple use will optimise resource utilisation and sustainability. This can be put in the form of a testable hypothesis: when the only purpose assigned to land is intensive agriculture, problems of erosion, water pollution, and pest outbreaks occur within decades. The alternative hypothesis is that if a landscape is designed for both agricultural production and sustainable ecological processes (wildlife conservation, closed nutrient cycles) the above problems are lessened or even avoided. At the other extreme many ecological problems may arise from large tracts of abandoned land, such as fire, pests, loss of species diversity and so on. (It is beyond the scope of this chapter to present the sort of precise tests for these hypotheses that were always on my mind when doing research on hedgerow network landscapes).

Hedgerow network landscapes offer us a good model of multiple purpose land use. Not only are hedgerows useful from the agricultural point of view, as shelter for crops and cattle or sources of wood, but they are also valuable habitats for wildlife and important for landscape amenity. These various aspects have been recently reviewed by Forman and Baudry (1984). The organising committee of the Workshop asked me to focus on hedgerows and wildlife.

Hedgrow network landscapes are common within EC countries from Denmark to Portugal (Flatres, 1959), but they are vanishing rapidly. (Pollard *et al.*, 1974; INRA etc, 1976; Barr *et al.*, 1986; Harms *et al.*, 1984). I will first present our findings at the hedgerow and hedgerow network levels and then make some comments on the design of such landscapes within the agricultural framework.

Wildlife in hedgerow network landscapes: a brief review

A considerable amount of work on species living in hedgerows has been done in Europe during the 1960s and 1970s (Pollard *et al.*, 1974; INRA etc, 1976). The review by Forman and Baudry (1984) included work done in North America. Since then research done within the framework of landscape ecology (Brandt and Agger, 1984; Forman and Godron, 1981, 1986; Berdoulay and Phipps, 1985) gave better insights into the various levels of ecological organisation involved and into the dynamics of the wildlife populations involved. Hedgerow structure and structure of the networks made of interconnected hedgerows are the two main features to which we have to pay attention. A striking point is that biological patterns are fairly similar in very different contexts (e.g. France and eastern North America) though hedgerow and landscape dynamics are different (Baudry, 1985).

Many interactions take place between hedgerows and adjacent fields, especially among insects (Pollard *et al.*, 1974; Lefeuvre *et al.*, 1976; Bowden and Dean, 1977). (It is still unclear whether or not those interactions are important in terms of pest control (Thresh, 1981)). Here, I shall concentrate on two questions: what are the relationships between hedgerow structure and species composition and how does landscape structure affect this species composition?

General concepts

Hedgerows are corridors (Forman and Godron, 1981); that is, landscape elements with a narrow width compared to their length. They are generally composed of three layers: herbs, shrubs and trees. The term hedge refers to a strip of shrubs trimmed periodically. In terms of habitat characteristics a corridor consists mainly of edges, while an interior habitat (with microclimatic conditions more or less independent of surrounding fields) may exist in wide corridors.

Hedgerows are usually interconnected into a network, and joined to woods, moorland and old fields (see Figure 11.1). The network can be described in terms of hedgerow density, number and types of connections (Baudry and Burel, 1985).

I use the concept of connectedness to refer to spatial relationships in structural patterns (Baudry, 1984), that is the number of connections among landscape elements. The concept of connectivity (Merriam, 1984) refers to functional interactions among individuals of the same species. Relationships between connectivity and connectedness are not always straightforward (Baudry and Merriam, 1987).

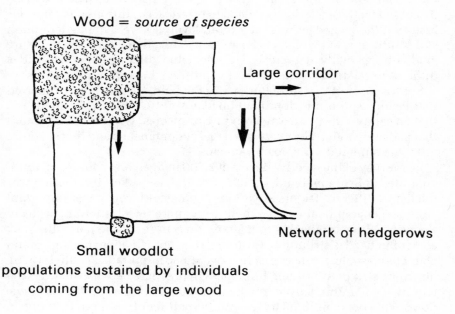

Figure 11.1   Hedgerow network and attached woods

Hedgerow characteristics and species composition

No species are restricted to hedgerows. Hedge flora and fauna derive from woodland, moorland, grassland or cropland but forest edge species predominate (Forman and Baudry, 1984). Hedgerows are important as habitat for species that cannot thrive in cultivated land, that is mainly woodland species, but also grassland species in cropland regions. Hedgerows may also be the only suitable habitat for a species in a specific region: Touffet (1976) reports that *Lithospermum diffusum* has the northern limit of its range in hedgerows in western Brittany. In a vegetation survey in north-eastern Brittany, Baudry and Burel (1978) found that uncultivated areas were restricted to poor soils (wet or shallow), while on loamy soils, hedgerows were the only refuges for plants.

Hedgerow vegetation

In the temperate zone, the most frequent tree species are *Quercus*, *Ulmus, Fraxinus* and *Populus. Crataegus, Prunus, Rosa* and *Rubus* are the most common shrubs. Apart from hedgerow structure, which will be discussed later, vegetation is related to physical conditions and land-use history. When a hedgerow is at ground level, it is very dependent upon soil conditions, but when it is on an earth bank, there is a steep moisture gradient especially if there is an associated ditch; in this case, it is difficult to relate vegetation to soil conditions in surrounding fields.

From a study of the Isle of Purbeck in England, Hooper (1976) reported that *Prunus spinosa* is dominant on clay soils, *Crataegus* on chalk and that on sandy soils there was a mixture of species. Willmot (1980) found that hedgerows along brooks were largely spontaneous and those around villages contained many exotic species.

In a survey of the woody species of 2,300 hedgerows in fourteen municipalities in Brittany, Baudry (1985) found that species distribution was related to different factors at various ecological levels. Most important were biogeographical factors: there were more forest interior shrubs in the wet, cold central areas of Brittany than in maritime parts. Bedrock also influenced distribution: *Castanea* was more frequent on granite and *Quercus* on shale. Farmers practices were also important, (Touffet, 1976), but, more strikingly, former differences in practices between districts led two adjacent municipalities on the same bedrock to have different hedge floras. This is exemplified by two municipalities of southern Brittany: in one of them an average of 8.3 woody species were found per hedgerow whilst in the other the average was only 5.9.

A study of species assemblages revealed 4 basic groups with plants originating from moorland, forest interior, wet woodland and forest edges. Plants from these groups could occur alone or combined. Some species (*Ulex europaeus, Erica cinerea*) were only found in hedgerows in former moorland cleared in the nineteenth century.

In the New Jersey hedgerows study by Baudry and Forman (unpublished data), there is a clear effect of hedgerow width on species composition (see Figure 11.2). This pattern was not found by Helliwell (1975) in narrow British hedgerows (3m). Unpublished data of Baudry and Burel in France show that tracks bordered by two hedgrows have more forest interior species on the shady side than single hedgerows do.

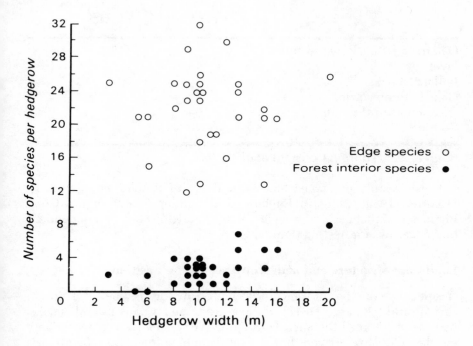

Figure 11.2   Hedgerow width and plant species composition

Source: Baudry and Forman, unpublished.

Hedgerow structure and birds

Birds are known to be sensitive to vegetation structure in any type of habitat (MacArthur and MacArthur, 1961). In a study of Brittany hedgerows, Constant *et al.* (1976) related bird density and species richness to both plant species composition and vegetation structure (see Table 11.1).

Table 11.1 Hedgerow structure and bird diversity

Type of hedgerow	Bird density (no. of pairs per km)	Species richness
Oak trees with a dense shrub layer	49	20
Pollarded oaks	46	15
Oaks with no shrubs	22	7
Ulex and *Sarothamnus*	34	13
Conifers	20	10

Source: derived from Constant *et al.* 1976.

These results are consistent with those of Pollard *et al.* (1974), Osborne (1984; 1985) in England and Yahner (1982) in the Great Plains shelterbelts. Game species such as partridge are also dependent on hedgerows for nesting (Rands, 1986).

Landscape structure and hedgerow species composition

A species may be present in a hedgerow or a wood lot in an agricultural landscape, because those are remnant patches and species has survived since the surrounding matrix was cleared. Alternatively species may have arrived through subsequent colonisation. Landscape structure may be critical for colonisation and population turnover. Landscape ecologists have carried out field studies and developed computer models to assess relationships between species composition of hedgerows and landscape structure (Baudry and Merriam, 1987). The general hypothesis is that interactions among landscape elements affect their species composition (Forman, 1981; Baudry, 1985).

Connections among landscape elements and plant species

The following figures are from unpublished data of Forman and Baudry collected in two hedgerow networks in New Jersey (USA). In Figure 11.3 the number of forest species in each hedgerow is plotted against the number of species in the richest connected element. It shows that, generally, a hedgerow does not have more species than its richest neighbour. Species richness of connected elements appears to be a strong constraint at the network level.

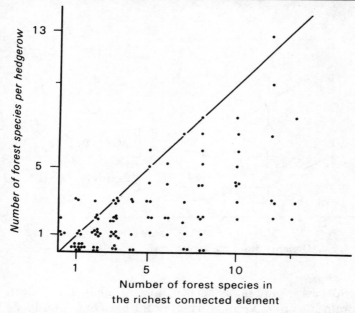

Figure 11.3 Effect of connections among hedgerows on species richness

Source: unpublished data from Baudry and Forman.

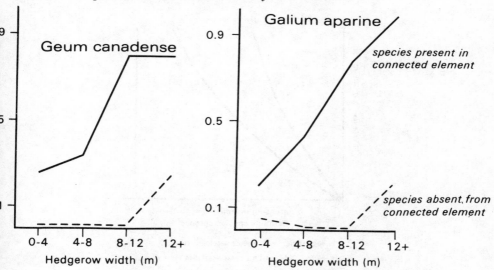

Figure 11.4 Connectedness and species frequency

Source: unpublished data from Baudry and Forman.

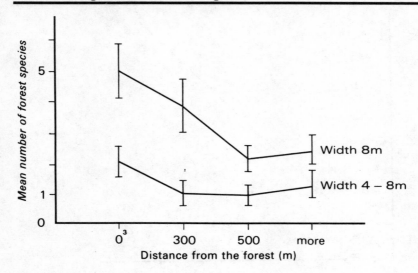

Figure 11.5 Species richness as a function of hedgerow width and distance from a source: verticla bars are 95 per cent confidence intervals

Source: unpublished data from Baudry and Forman.

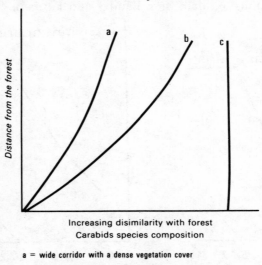

a = wide corridor with a dense vegetation cover

b = narrow corridor with a dense vegetation cover

c = narrow corridor with poor vegetation

Figure 11.6 Hedgerow structure, distance to a forest and Carabids
Source: adapted from Burel, unpublished.

In Figure 11.4 the frequency of two species in hedgerows is presented as a function of hedgerow width and presence of the species in connected hedgerows. In wide hedgerows (potentially suitable habitat), the two species are frequent only if present in adjacent hedgerows (or woods). Distance from woods is also an important factor in these spontaneous hedgerows (Figure 11.5), that is hedgerows that result from bird dispersed seeds, e.g. along fence lines. Patterns found in France are similar, though less clear cut because of historical disturbance at the hedgerow level (Baudry, 1985).

Landscape structure and Carabidae populations

Burel (1987 and unpublished data) studied the distrubution of ground beetles in an agricultural landscape attached to a forest. She was examining the role of hedgerows as corridors for forest species. Her major findings were that species of forest Carabids (e.g. *Abax ovalis*, *Platysma nigrum*) are found kilometres away from the forest in tracks bordered by two hedgerows, and also in connected hedgerows with a dense canopy. These species certainly reproduce in hedgerows. Some species (e.g. *Archicarabus nemoralis*) are not found far from the forest, so species composition similarity between hedgerows and forest decreases as distance increases. There are also some species restricted to the forest. Figure 11.6 shows broad patterns. Those appear to be similar to the ones for plants with strong interactions between hedgerow structure and distance from a source of forest species.

Landscape structure and birds

Both species richness and density are higher in landscapes with many hedgerows as shown by data from Constant *et al.* (1976) (see Table 11.2). A dense network of hedgerows supports species that nest in the shrub layer and in dead trees. As hedgerow density decreases open field species (*Alauda arvensis*) colonise the landscape.

Table 11.2 Hedgerow density and birds

	Dense hedgerow network	Open fields
Bird density (no. of pairs per 100 ha)	99	35.3
Species richness	40	23

Source: Constant *et al.* 1976.

Landscape structure and small mammals

Merriam and co-workers (Merriam, 1984; Middleton and Merriam, 1981; 1983; Wegner and Merriam, 1979) have an important project on small mammal populations being carried out in eastern Ontario (Canada). The frequent local extinctions of populations of small mammals during winter are rapidly counter-balanced in spring by migrants from other patches of wood. This colonisation process (connectivity among woodlots) is insured by hedgerows which connect woods. Computer models (Lefkovitch and Fahrig, 1985) show that not only the existence of connections is important but also their number. When a woodlot is connected to three or four others instead of one there are several sources of individuals.

Conclusion

From the few results presented here, it is clear that if we want to protect hedgerows for wildlife conservation, we must study interactions within the whole landscape, at the same time as we study hedgerow structure. The view that hedgerows are closed ecological systems with independent dynamics (within a succession process, for example) is not tenable. Landscape constitutes the appropriate scale of observation for wildlife management (see O'Neill *et al.*, 1986, for the problem of scale). Conservation of isolated hedgerows will probably not sustain many ecological processes in the long run. Landscape dynamics, shaped by farming activity, is a matter of concern.

Landscape design within the agricultural framework

It was certainly not the purpose of ancient farmers to design and manage hedgerow networks for wildlife, though they did collect berries. Ecological processes we are dealing with are self-organising ones, within the structures designed by human activities in interaction with physical conditions. Most of the time farmers and land owners are solely responsible for their hedgerows. In France, exceptions are during land consolidation operations where an ecological survey is mandatory (Baudry and Burel, 1984; Burel, 1984). In Britain some farms are managed for both agriculture and wildlife (Leonard and Cobham, 1977) and the management agreement may save some valuable habitats. As agricultural policies are (or will be or should be) shifting emphasis from production to land management, policy makers need tools to evaluate those changes in terms of environmental impacts.

For a farmer the basic question may be 'How do I make a hedgerow pay?'; it can be either by its own production or positive influence on crops, cattle, soil conservation (though it is unlikely that long-term benefits can be taken into account by economists) or by being a source of funding from any organisation willing to protect hedgerows. I did research and worked as a planner from the first standpoint, looking at the place of hedgerows and associated ditches in agricultural systems and global landscape functioning.

As costs of agricultural inputs increase and prices of outputs level off or decrease, consciousness of the importance of hedgerows increases among farmers; but labour intensive management practices are a deterrent for many people. From these observations, I suggest further research in two directions: first, within a physical landscape pattern (soil, slope, climate) we should look for alternative hedgerow network designs helpful for the (present) agricultural systems. This type of research would also help to develop sustainable agricultural systems (Altieri *et al.*, 1983; Lowrance *et al.*, 1984).Second, it is necessary to have hedgerows easy to manage, with different structures for different purposes (a good windbreak does not have the same structure as a shelterbelt design to protect a road from snow accumulation).

Compatibility between design for agriculture and wildlife may not be out of reach. If we look at a hedgerow network from a farmer's standpoint, we see that to be efficient as a windbreak, the network must be continuous, as must be the network of ditches for drainage or banks for erosion control. So connectedness will be a consequence of agricultural efficiency of the network.

Conclusion

In man-dominated landscapes the presence of hedgerows may be critical for many species as permanent or temporary habitats and as corridor to move across the landscape between suitable patches. This type of landscape structure certainly maintains species composition of woodlots close to that of large forests (Middleton and Merriam, 1983). As agricultural systems and land-use patterns change, it is important to think of hedgerows as parts of those systems.

References

Altieri, M.A., Letourneau, D.K. and Davis, J.R. 1983. Developing sustainable agroecosystems. *BioScience*, *33*, 45–49.
Barr, C., Benefield, C., Bunce, B., Riddsdale H. and Whittaker, M. 1986. *Landscape Changes in Britain*. Institute of Terrestrial Ecology, Monks Wood, Huntingdon.
Baudry, J. 1984. Effects of landscape structure on biological commu-

nities: the case of hedgerow network landscapes. In Brandt, J. and Agger, P. eds, *Methodology in Landscape, Ecological Research and Planning*. Vol. 1. Theme 1: Landscape ecological concepts. Roskilde University Center. Denmark. 55–65.

Baudry, J. 1985. Utilisation des concepts de Landscape Ecology pour l'analyse de l'espace rural: occupation du sol et bocage. Thèse de Doctorat d'état. Université de Rennes. 487 p.

Baudry, J. and Burel, F. 1978. Contribution à la connaissance écologique du bassin versant de la Ranie. Thèses de 3eme cycle. Universife de Rennes.

Baudry, J. and Burel, F. 1984. Landscape project: 'Remembrement': Landscape consolidation in France. *Landscape Planning, 11,* 235–41.

Baudry, J. and Burel, F. 1985. Systeme ecologique, espace et therie de l'information. In Berdoulay,V. and Phipps, M., eds, *Paysage et système,* 87–102.

Baudry, J. and Merriam, H.G. 1987. Connectivity and connectedness: functional versus structural patterns in landscape. Communication au 2nd Seminar of IALE (Munster) — Allemagne.

Berdoulay, V. and Phipps, M. 1985. *Paysage et système.* Editions de l'université d'Ottowa. 195 p.

Bowden, J. and Dean, G.J.W. 1977. The Distribution of Flying Insects in and near Tall Hedgerows. *Journal of Applied Ecology,* 14, 343, 354.

Brandt, J. and Agger, P., eds, 1984. *Methodology in Landscape Ecological Research and Planning.* Roskilde University centre, 5 vol., 118, 150, 153, 171, 235.

Burel, F. 1984. Use of landscape ecology for the management of rural hedgerow network areas in Western France. In Brandt, J. and Agger, P. (eds). *Methodology in Landscape Ecoplogical Research and Planning* Vol. 2, Theme 2: Methodology and techniques of inventory and survey; 73–81.

Burel, F. 1987. Biological patterns and structural patterns in agricultural landscapes, communication at 2nd International seminar of the International Association for Landscape Ecology.

Constant, P., Eybert, M.C. and Maheo, R. 1976. Avifaune reproductrice du bocage de l'Ouest. In INRA, CNRS, ENSA et Université de Rennes, *Les Bocages: Histoire, Ecologie, Economie,* 327–32.

Fahrig, L., Lefkovitch, L.P. and Merriam, H.G. 1983. Population stability in a patchy environment, in Lauenroth, W.K., Skogerboe, G.V. and Flug, M. eds, *Analysis of Ecological Systems: State of the Art in Ecological Modeling,* Elsevier, New York, 61–7.

Fahrig, L. and Merriam, G. 1986. Habitat patch connectivity and population survival. *Ecology, 67,* 1762–8.

Flatres, P. 1959. Les structures rurales de la frange atlantique de l'Europe, *Géographie et Histoire Agraire,* Mémoire no 21 Annales de l'Est, 193, 202.

Forman, R.T.T. 1981. Interaction among landscape elements: a core of landscape ecology. In Thallingii, S.P. and de Veer, A.A. eds, *Perspectives in Landscape Ecology,* Pudoc, Wageningen, 57–64.

Forman, R.T.T. and Baudry,J. 1984. Hedgerows and hedgerow networks in landscape ecology, *Environmental Management, 8,* 499–510.

Forman, R.T.T. and Godron, M. 1981. Patches and structural components for a landscape ecology, *Bio Science, 31,* 733–740.

Forman, R.T.T. and Godron, M. 1986. *Landscape Ecology,* Wiley, New York, 619 p.

Harms, W.B., Stortelder, A.H.F. and Vos, W. 1984. Effects of intensification of agriculture on nature and landscape in the Netherlands, *Ekologie (CSSR), 3,* 281, 304.

Helliwell, D.R. 1975. The distribution of woodland plant species in some Shropshire hedgerows, *Biological Conservation, 7,* 61–72.

Hooper, M.D. 1976. Historical and biological studies on English hedges, in INRA, CNRS, ENSA et Universite/ de Rennes, *Les Bocages: Histoire, Ecologie, Economie,* 225,227.

INRA, CNRS, ENSA and Université de Rennes (1976) *Les Bocages: Histoire, Ecologie, Economie,* p. 586.

Lefeuvre, J.C., Missonnier, J.and Robert, Y. 1976. Characterisation zoologique. Ecologie animale (des bocages), Rapport de synthèse.

Lefkovitch, L.P. and Fahrig, L. 1985. Spatial characteristics of habitat patches and population survival *Ecological Modelling, 30,* 297, 308.

Leonard, P.L. and Cobham, R.O. 1977. The farming landscape of England and Wales: a changing scene, *Landscape Planning, 4,* 205–36.

Lowrance, R., Stinner, B.R. and House, G.J. 1984. *Agricultural Ecosystems.* Wiley, New York. 233 p.

MacArthur, R. and MacArthur, J. 1961. On bird species diversity, *Ecology, 42,* 597, 598.

Merriam, H.G. 1984. Connectivity: a fundamental characteristic of landscape pattern. In Brandt, J. and Agger, P. eds, *Methodology in Landscape Ecological Research and Planning,* Vol. 1, Theme 1, landscape ecological concepts, Roskilde University centre, Denmark, 5–15.

Middleton, J. and Merriam, H.G. 1983. Distribution of woodland species in farmland woods, *Journal of Applied Ecology, 20,* 625–44.

Middleton, J. and Merriam, G. 1981. Woodland mice in a farmland mosaic. *Journal of Applied Ecology*, *18*, 703–10.

O'Neill, R.V., de Angelis, D.L. Walde, J.B. and Allen, T.F.H. 1986. *A Hierarchical Concept of Ecosystems*, Princeton University Press, Princeton, NJ, 253 p.

Osborne, P. 1984. Bird numbers and habitat characteristics in farmland hedgerows. *Journal of Applied Ecology*, *21*, 63, 82.

Osborne, P. 1985. Some effects of dutch elm disease on the birds of a Dorset dairy farm. *Journal of Applied Ecology*, *22*, 681, 691.

Peterken, G.F. and Game, M. 1981. Historical factors affecting the distribution of *Mercurialis perennis* in Central Lincolnshire, *Journal of Ecology*, *69*, 781–96.

Pollard, E., Hooper, M.D. and Moore, N.W. 1974. *Hedges*, W. Collins and Sons, London, 256 p.

Rackham, O. 1986. *The History of the Countryside*, J.M. Dent, London, 445 p.

Rands, M.R.W. 1986. Effect of hedgerows characteristics on partridge breeding densities. *Journal of Applied Ecology*, *23*, 479, 487.

Risser, P.G., Karr, J.R. and Forman, R.T.T. 1983. Landscape ecology directions and approaches, The Illinois Natural History Survey, Natural Resources Building, 607 East Peabody Drive, Champaign, Illinois 61820, 16 p.

Thresh, J.M. (ed). 1981. *Pest, Pathogens and Vegetation*. Pitman, New York, 517 p.

Touffet, J. 1976. Caracterisation botanique des haies et des talus, in INRA, CNRS, ENSA et Université de Rennes, *Les Bocages: Histoire, Ecologie, Economie*, 211, 217.

Wegner, J.F. and Merriam, G. 1979. Movements by birds and small mammals between a wood and adjoining farmland habitats. *Journal of Applied Ecology*, *16*, 349–58.

Willmot, A. 1980. The woody species of hedges with special reference to age in Church Broughton Parish, Derbyshire, *Journal of Ecology*, *68*, 269,285.

Yahner, R.H. 1982. Avian use of vertical strata and plantings in farmstead shelterbelts. *Journal of Wildlife Management*, *46*, 50–60.

12 | The status of hedgerow field margins in Ireland

R.Webb, An Foras Forbatha, Dublin

Hedgerows are relatively recent additions to the Irish landscape. Most hedges were planted after the 1667 Cattle Act and subsequent Enclosure Acts. Enclosure was not popular with farmers who had been used to grazing common land without boundaries but, over time, hedges became an integral part of the 'natural' environment.

Ireland is the least wooded country in the European Community, with forests covering around 5 per cent of the land area, and with broadleaved high forest covering about 0.5 per cent. With the continued destruction and little replanting of these woods, hedgerows take on an added importance as substitute woodland habitats for much of Ireland's wildlife. Nearly two-thirds of Ireland's bird species nest in hedges, and the importance of hedgerows as a wildlife habitat can be seen when the possible area of hedgerows in the country is calculated. Given that there is an average of 0.022 km2 of hedge per km2 and that approximately 49,000 km2 of land under crops and pasture may support hedges, the total area of hedgerow habitat in Ireland may be in the order of 1,078 km2, or 1.5 per cent of the country. This is three times the area covered by deciduous high forest, or five times the combined area of Ireland's four national parks (Webb, 1985).

However, few studies to date have concentrated on the current status of hedgerows in Ireland from the point of view of wildlife or landscape. O'Sullivan and Moore (1979) mapped the composition of field boundaries in 1974 from 2,000 site records. A number of local studies have examined hedgerows in detail in particular areas: Synott (1975) studied hedgerow dating in part of Co. Meath and found that dating techniques developed in England (Pollard *et al*, 1974) did not apply to the area studied. Other studies include Feehan

(1983) on field patterns in Co. Laois, Hickie (1985) in Co. Dublin, and an unpublished study by the Macroom District Environmental Group, which examined 30 km2 in Cork for landscape change and hedgerow loss.

Only two studies have concentrated on hedgerows on a nationwide basis. The first study was carried out as part of a report into the state of the environment (An Foras Forbartha, Cabot, 1985), involving a preliminary investigation of hedgerow removal and hedgerow trees, and in 1985 the Irish Wildlife Federation launched a national survey of Irish hedgerows with the specific aim of collecting more information on this important habitat (IWF 1987). The purpose of this chapter is to outline the findings of these two studies.

Hedgerow structure

The IWF survey distributed around 8,000 survey cards to members in all parts of the country, including to 800 second-level schools. Approximately 4 per cent were returned. Completed surveys covered hedges in 23 counties. Over 8 per cent of all surveys were completed by school children.

The majority of surveyed hedges consisted of thorny shrubs and trees on an earth or stone bank, with a ditch or drain running alongside. A small number (3 per cent) had no bank. Variation in hedge height and width was mainly a reflection of management practices.

Mechanically trimmed hedges were generally less than 2 m in height, while unmanaged and overgrown hedges reached a height of up to 5 m. In general, however, Irish hedges averaged 3 m in height. Hedge width varied from 1 m to 4 m, but most were under 2 m, with an average width of 1.8 m. The height of the bank was less variable and ranged from 0.3 to 2.5 m with an average height of 0.9 m.

Most hedges (68 per cent) had some type of associated ditch or drain. This structure was probably originally built to improve drainage on wet soils and provided the material for the bank. At present, 52 per cent of these ditches hold water, but those which are dry are so because of the increasing use of land drainage pipes.

Flora

Around thirty-seven species of shrubs and trees were recorded in Irish hedgerows during the survey. Many factors influence the species composition of hedges such as soil type, drainage, climate, shade and planting policy. Thorny bushes are the usual component of hedges, the most important being hawthorn (*Crataegus monogyna*) and blackthorn (*Prunus spinosa*), the latter present as a coloniser rather than through planting. Gorse (*Ulex sp.*) is dominant on infertile upland soils, especially in parts of counties Wicklow, Donegal and Kerry. Willow (*Salix sp.*) and alder (*Alnus glutinosa*) are common on poorly drained soils such as in the north-west, Cork and Kerry. The introduced shrub, *Fuschia*, is characteristic of hedges in Kerry, West Cork and parts of Donegal. Holly (*Ilex europaeus*) is found throughout the country, while beech (*Fagus sylvatica*) is common in the east and midlands, and is usually associated with estate plantings.

A total of 105 wild flowers was recorded in the surveyed hedges. The most common were primrose (*Primula vulgaris*), violet (*Viola riviniana*), nettle (*Urtica dioica*), foxglove (*Digitalis purpurea*), and Herb robert (*Geranium robertianum*), with bluebell (*Endymion non-scriptus*) and Dog's mercury (*Mercurialis perennis*), indicative of ancient hedgerows.

Among the rarer occurrences were Irish Spurge (*Euphorbia hyberna*), and wood sage (*Teucrium scorodonia*) from Co. Kerry, Wild thyme (*Thymus praecox*) from Co. Laois, flax (*Linum sp.*) from Co. Wexford, and marjoram (*Origanum vulgare*) in Co. Kilkenny. The Lady fern (*Athyrium felix-femina*) was only recorded from hedges in Co.Kerry.

Hedge removal

In the An Foras Forbartha study, over thirty randomly selected one-kilometre squares were examined around the country. The lengths of hedgerows were measured from the 1936 Ordnance Survey 6-inch scale maps, which were compared with the 1973 air photographs available from the Geological Survey of Ireland. The present situation was then assessed in the field. Work will be continuing, but preliminary results suggest an average hedgerow loss of around 16 per cent since 1938, which parallels that in the Macroom study, compared with an average of around 30 per cent in Britain (O'Connor and Shrubb, 1986). In some areas there seems to have been a net gain in hedgerows due to land reorganisation. Hedge removal appears to be localised, with more removal taking place on the larger farms, irrespective of whether they are tillage or grazing farms. Areas of greatest clearance seem to be in the south Laois/south Kildare area, with about a 30 per cent removal of hedgerows.

Hedges were originally planted as purely functional parts of the farm, designed to restrict stock movement and act as farm boundaries. Since the advent of modern farming methods throughout Europe, hedges have often become surplus to requirements. The arguments against them include restriction on field size and machinery, shading of crops, harbouring of pests and weeds, and expense and time in maintenance, but these ignore possible agronomic benefits (see other chapters in this part).

Hedgerow trees

In order to obtain a possible guide to the status of hedgerow trees, an examination of the trees in the thirty km squares was carried out. A total of 1,185 trees was distributed among the four age classes of 'sapling', 'young', 'mature', and 'overmature' (see Table 12.1):

Table 12.1   Age structure of hedgerow trees in thirty 1 km squares

| Species | Age class | | | | %occurrence From total number sampled |
	Sapling	Young	Mature	Over-mature	
Fraxinus	312	216	99	3	53%
Fagus	30	53	101	1	15%
Ulnus	40	38	42	6	10%
Acer pseudoplatanus	10	63	25	1	8%
Quercus	8	17	17	3	5%
Others*	41	36	12	1	76%
Total	441	423	306	15	1,185

Source: R. Webb Field Survey, 1984
*Prunus, Salix, Sorbus, Ilex, Aesculus

To ensure replacement of trees and to allow for mortality, the numbers of saplings should about equal the number of trees in all other classes. As this is not the case in the survey areas (the situation being broadly similar to England twenty years ago - Forestry Commission 1965), even if no more trees were felled, the total number will decrease. It is, however, difficult to assess the numbers of ash saplings in tall hedges, so the ratio may be more favourable. As there are more saplings than semi-mature trees, and both classes equal the numbers of mature trees, the situation at first sight is not too discouraging.

It is worth noting, however, the complete dominance of ash (Fraxinus) in all age classes, and the lack of saplings of all other species. The dominance of Fraxinus was also noted in the IWF survey. At 53 per cent, ash forms a major hedgerow tree, and saplings are vulnerable to mechanical cutting. It is not a very long-lived tree and being shallow rooted is also vulnerable to cultivation.￿

Beech (*Fagus*), while not a major hedgerow tree, was planted in vast numbers during the last century as an amenity tree and for shelter. It has become an essential element in the Irish landscape, yet there are no saplings to replace the trees now dying of old age in increasing numbers. Over half of the trees damaged in the gales of 1984 were over-mature beech (Webb, 1986). Oak (*Quercus*), accounts for only 5 per cent of hedgerow trees and may be said to be uncommon in the countryside, outside of our native woods. In areas such as the north and east of the country, where *Quercus* once predominated, Sycamore (*Acer pseudoplatanus*) appears to have filled its position in the hedgerows. This may result in somewhat impoverished fauna in view of the fact that *Quercus* supports a far greater variety of wildlife than Acer (Southwood, 1961).

The elm (*Ulnus sp.*) has virtually disappeared as a tree from Irish hedges following the introduction of the aggressive strain of Dutch elm disease (*Ceratocystis ulmi*) around 1976. The disease has caused great devastation in counties Limerick and Dublin, where elm was a major constituent of the hedgerow. In other areas it is not so important in hedges, as in England. While occasional healthy trees may be found, most of those surveyed showed signs of the disease. The species itself will continue to be present, but only as saplings.

Management

Management practices have a direct bearing on the long-term ability of a hedge to act as a stock-proof barrier as well as affecting its wild-life potential. In the past, the maintenance of hedges through laying or hand clipping was part of the annual farm duties. However, even the mechanical cutting of hedges is expensive. Hiring a contractor can cost £12–£16 per hour. The bill for a 50 ha farm could come to £300 or more every two years, depending on the number and condition of the hedges (Hickie, 1986). Results of the IWF survey indicated that 82 per cent of Irish hedgerows surveyed had not been recently managed and were overgrown. A total of 16 per cent, mostly roadside hedges, were mechanically trimmed flat-topped structures, where few saplings were allowed to grow into mature trees. Roadside hedges are often trimmed by the local highway authority, and although by law hedges should not be cut between April and September, this is often ignored.

Only 2 per cent of surveyed hedges were hand-clipped, while one hedge in Co. Louth was reported to have been laid in the traditional manner. Lack of maintenance has resulted in 34 per cent of hedges becoming 'leggy' and requiring wire as reinforcement. Approximately 20 per cent of hedges were described as being reasonably stock-proof; 40 per cent good, while only 6 per cent were very good.

Conclusions

Although the two surveys are limited in the number of hedges surveyed, they do highlight some important considerations. Hedgerows are undoubtedly important wildlife refuges, especially for woodland flora and fauna. The decline in hedgerow trees, however, especially oak (*Quercus*) gives cause for concern. Good colonisers, such as *Fraxinus* and *Acer pseudoplatanus*, are now the dominant hedgerow trees. In most mechanically trimmed hedges even these trees do not grow to maturity.

With regard to management, the lack of maintenance, which allows hedges to become too tall and gappy, not only threatens their wildlife value but also their ability to be used as stock-proof barriers. Ironically, economic factors may well be responsible for so many hedges remaining in the countryside. Hedge removal is expensive — anything up to £5 per metre and more if trees are present. Grants for this purpose are no longer as favourable as they were. Also, grass and crops growth will usually be poor on the land so reclaimed, for several years.

The main reason that we have not seen the wholesale removal of hedges is that Ireland continues to have a high proportion of dry-stock and dairy enterprises. Hedges do not interfere with these operations nearly as much as with large-scale tillage, and they are often valued by farmers for the shelter they afford livestock. We shall continue to see some hedge removal in areas where fields are still small, such as in Connaught and the south-west, but it is unlikely that there will be wide-scale hedge removal in future, but this is no reason to be complacent. In particular, practical advice needs to be available to farmers regarding hedgerow management methods which are sympathetic to wildlife, as well as incentives for broadleaved tree planting in hedges.

References

Cabot, D. ed. 1985. *The State of the Environment.* An Foras Forbartha, Dublin.

Feehan, J. 1983. *Laois: An Environmental History.* Ballykilcavan Press 551p.

Forestry Commission of Great Britain. 1965. *Census of Woodlands.* HMSO.

Hickie, D. 1985. A hedge study in north Co. Dublin. Unpublished MSc thesis in Environmental Science, Trinity College, Dublin.

Hickie, D. 1986. Hedges: important wildlife refuges, in *Irish Shooting Companion.* National Association of Regional Game Councils, 81–7. Irish Wildlife Federation. 1987. Irish Hedgerow Survey. Report No. 1 in *The Badger,* No. 27.

O'Connor, R.J. and Shrubb, M. 1986. *Farming and Birds,* Cambridge University Press, Cambridge.

O'Sullivan, A.M. and Moore, S.J., J.J. 1979. Composition of Field Boundaries 1974, in *Atlas of Ireland.* Royal Irish Academy, Dublin.

Pollard, E., Hooper, M.D., and Moore, N.W. 1974. *Hedges.* New Naturalist Series, Collins.

Southwood, T.R.E. 1961. The number of species of insect associated with various trees. *Journal of Animal Ecology,* No.30.

Synott, D. 1975. Hedge dating in Ireland. *Farm Bulletin. May.*

Webb, R. 1985. Farming and the landscape, in *The Future of the Irish Rural Landscape.* ed. Aalen, F.H.A., Trinity College, Dublin, 80–92.

Webb, R. 1986. Results of a survey into gale damage to roadside trees, in *Irish Journal of Environmental Science,* 3. 38–40.

13 Agricultural changes in scrub and grassland habitats in Europe

P. *Devillers*, Institut Royal des
Sciences Naturelles de Belgique, Brussels.

The fauna of Europe can be conveniently divided into five very broad groups with natural foraging preferences in, respectively:

— forest tree-tops;
— forest edge mantle, recolonisation and degradation stages;
— dry, open, herbaceous areas;
— aquatic habitats;
— rocky habitats.

Forest tree-top faunas are potentially dominant in most of Europe and must indeed have been overwhelmingly represented during the short time-span between the reinstallation of forest after the last glaciation and the onset of its destruction by human activity.

The second and third groups could only have occupied very restricted edaphic, microclimatic or transient enclaves in a pre-neolithic environment and their centres of abundance and differentiation may have been, particularly in the case of grassland species, entirely outside the region. Both groups profited considerably from the agro-pastoral activities of man. At present, among breeding birds of the Community, for instance, species that forage preferentially in pre-forest and forest edge habitats represent 38 per cent of the total non-aquatic fauna, those that forage essentially in open grassy areas 26 per cent. For both groups, future survival in our territory is closely linked with the evolution of agricultural practices. The intrinsic characteristics of the two groups are different. The birds of scrubby and mixed habitats mostly include

---

See also Chapter 3, 'Belgium'.

adaptable species of broad ecological tolerance, those of grassy areas mostly very demanding specialists. Thus, Annex 1 of the Bird Directive (79–409), which lists species highly vulnerable to alterations of their habitat and comprises about 35 per cent of the total breeding avifauna, includes only 16 per cent of the birds of scrub habitats but 48 per cent of those of grassy habitats making these the most threatened ecological group, ahead of even aquatic species, with 43 per cent.

The situation of the scrub and edge group could, however, evolve rapidly. The most important types of habitats for its representatives have been:

(a) the Mediterranean scrubs, sufficiently ancient and extensive to have acquired complex communities and even harboured entire radiations of upper organisms, of which the most spectacular is probably that of the insect-mimicking orchid genus *Ophrys*; these scrubs are now rapidly disappearing through urbanisation or closing up by forest recolonisation following the widespread disaffection for pastoralism in Mediterranean countries, except Spain;

(b) the Atlantic heaths, much reduced by processes that are similar but have been operating for a longer time;

(c) the *Prunetalia* brushes of deciduous forest lands, usually among the first victims of urbanisation, agricultural rationalisation, organisation of green spaces, forestry intensification;

(d) more directly linked to this chapter's preoccupations, agricultural landscapes of mixed ground cover, such as English parkland, Spanish *dehesa*, French or Belgian *bocage*, discussed in Professor Noirfalise's Chapter 3: the *bocage*, a reticulated network of pastures, hay-meadows, hedgerows and small woodlots, is extremely favourable to the enthomofauna, to small mammals and to insectivorous birds, including vulnerable specialists such as shrikes (*Lanius*) or wryneck (*Jynx torquilla*); formerly covering surfaces large enough for the perpetuation of complex faunas, it has undergone a considerable regression, evaluated by Professor Noirfalise at 75 per cent to 80 per cent for high Belgium, as a result mostly of plot reallocation and agri-cultural intensification: the Spanish *dehesa*, which combines cereal cultivation and livestock grazing in shade with acorn harvesting, has an exceptionally rich fauna, including essential populations of large predators such as the endangered Pardel lynx (*Lynx pardina*) and Spanish imperial eagle (*Aquila heliaca adalberti*); it harbours most of the European wintering populations of the crane (*Grus grus*) its rapid reduction is the result of the abandonment of pig raising.

Turning now to dry grassland species, essentially of eastern and southern origin, their stronghold in Europe has long been areas of extensive sheep grazing on the one hand, particularly the sub-Mediterranean, supra-Mediterranean or Mediterranean calcareous grasslands and pseudosteppes, and regions of not too intensive cereal cultivation on the other hand. Among birds, bustards (*Otis tetrax*), sandgrouses (*Pterocles*), stone curlew (*Burhinus oedicnemus*), harriers (*Circus*), larks (*Alaudidae*), partridges (*Perdix alectoris*) and quail (*Coturnix coturnix*) are particularly representative of the group. All have suffered considerable decreases in number and reduction in range during the past century. A few examples: the Great bustard (*Otis tarda*) has disappeared from England and France, the Little bustard (*Tetrax tetrax*) has almost deserted the northern half of its range in France, the Short-toed lark (*Calandrella brachydactyla*), breeding not long ago to the Belgian border, has now retreated to the Mediterranean region, and the Pintailed sandgrouse (*Pterocles alchata*), precariously holding on in France, has abandoned many former haunts in northern Spain. The major causes of this phenomenom are the severe reduction in extensive sheep grazing, the constantly more widespread use of pesticides and herbicides, the conversion into intensive cultivation of marginal lands, mechanisation and earlier date of harvesting. Although this process is to a large extent irreversible, a set of reasonably simple measures may alleviate some of its effects, at least for selected species. Thus, direct protection of nests of Montagu's harrier (*Circus pygargus*) at the time of harvest is enabling the species to survive in some regions of intensified agriculture. Most effective is the preservation of strips or patches of semi-natural vegetation at the edges of fields, in less manageable areas such as triangular corners, and on marginal soils, notably the upper part of small hills in the Mediterranean region or the less well-drained areas in other climatic areas. These provide suitable habitats for species incapable of using the fields themselves — the Pintailed sandgrouse is one — and, perhaps even more significantly, provide food and shelter during and after the harvest for species that succeed in breeding in the fields, provided the young are old enough to move to these areas at the time of harvest.

Such measures are, together with the setting aside of sufficiently large semi-natural ecosystems, essential to the preservation of a rapidly dwindling fauna for which there may be little alternative. One might indeed feel unconcerned by its fate in the agricultural landscape, relying on the argument that, since the fauna has progressed thanks to traditional human activities, it can safely, once these activities cease to be favourable, be allowed to retreat to its range and habitats of origin. This, however, has ceased to be possible as natural steppe has all but vanished to the east of us, just as it has in North America. Thus, the steppe fauna has nowhere to retreat to and will survive in its acquired cultivation habitat or vanish. An example: the Spanish population of the Little bustard exceeds by several orders of magnitude that of the entire Soviet Union.

Finally, it should be noted that the measures needed are not necessarily an economic burden. In Belgium, for instance, the Grey partridge, an important game species, is in a critical situation. Management that would improve its status could certainly bring a substantial yield in hunting revenues. In the Mancha area of Spain, the Jose Maria Blanc Foundation, to whom I am grateful for this information, has demonstrated that a combination of disuse of pesticides, preservation of semi-natural vegetation at field edges and on marginal soils, and slight delay in harvesting time produced combined agricultural and hunting revenues exceeding those obtained in more intensively cultivated neighbouring properties while additionally maintaining reservoirs of vulnerable non-game species such as sandgrouses, bustards and eagles (*Aquila hieraaetus*).

14 | The dispersal of plants from field margins

E.J.P. Marshall, University of Bristol

In lowland Britain the field boundary, which is typically a thorn hedge, is a prominent part of farmland. Concern over the effects of intensive agriculture on wildlife has brought attention to this feature of the landscape. Where other habitats have been eroded, the field boundary may act as a reservoir and refuge for many plant, animal and bird species. Nevertheless, with less stock-rearing and more arable farming, the role of the hedge has been a source of debate for some 140 years. In particular, the impact of field boundaries on weed, pest and disease control has occupied farmer's minds. Changes in the management of the boundary and the adjacent crop have affected the composition of field boundaries, often with unforeseen results. In this chapter, studies of plant distribution patterns in arable crops are described and the implications for agriculture and the environment are discussed

Definitions

At a recent meeting organised by the British Crop Protection Council, the terms used to describe field margins were proposed (Greaves and Marshall, 1987) (see Figure 14.1). The main structure of the margin, such as a hedge, ditch, fence or wall, and the associated herbaceous vegetation is termed the boundary. The area between the boundary and the crop is termed the boundary strip and may include a barrier strip of grass or bare ground; this may not be present if the land is cultivated up to the boundary. The crop edge is self-explanatory. It was proposed that the term headland, used to define the area in which farm machinery turns, was not appropriate for the crop edge. However, this term is widely used to mean the area of the crop between the boundary strip and the first tractor-wheeling.

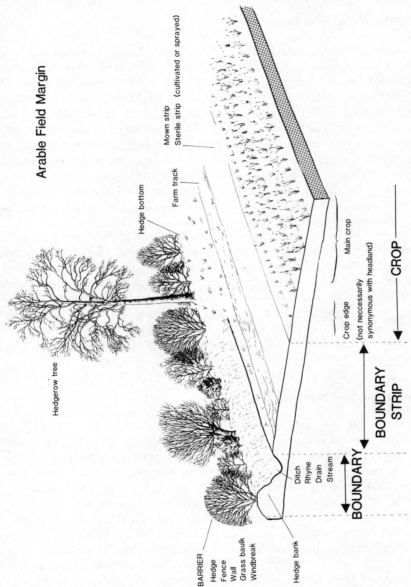

Figure 14.1   The principal components of an arable field margin

Source:

The role of the field boundary

Where animals are raised, the field boundary has an agricultural role of stock impoundment as well as defining an edge to the field. In arable land the boundary may have no major agricultural role, save that of defining an edge. However, recent research has been seeking to understand the role of the boundary in harbouring both beneficial and pest organisms (see Wratten, Chapter 15). In terms of the flora, about 500 higher plant species have been recorded from hedgerows in the United Kingdom. Potentially, such sites are diverse habitats, which may affect the maintenance of local populations of both plant and animal species. In terms of landscape, field boundaries undoubtedly contribute to visual diversity. The loss of hedges from parts of the eastern counties of England has been deplored by many environmentalists.

The requirements for the field boundary are that it should provide an edge, contain stock or keep trespassers out and that it should not interfere with the adjacent crop. Wildlife, game and landscape interests may be satisfied by providing a diversity of types of boundary which themselves contain a diversity of plants and animals. These requirements may be satisfied by a range of different boundary structures where particular interests have priority. The point of conflict is the potential interference by the boundary with the crop, particularly by the spread of weeds, pests and diseases.

Farming perceptions

It has long been a perception that field weeds spread from field edges into the crop. Interviews with 160 cereal growers at the 1985 Royal Agricultural Society of England Show, revealed that 30 per cent of farmers manage the edge of the crop or boundary in order to control weeds (Marshall and Smith, 1987). At cereal harvest, crop edges are often less ripe than the main crop and may contain green weed material. Both factors do not facilitate harvesting. The belief that weeds spread from the hedge has led a number of farmers to attempt to eliminate particular species in the herbaceous vegetation at the base of the hedge. Commonly, herbicides have been used to achieve this. If the perennial ground flora is controlled, opportunities for the germination of annual species are increased. It is therefore a common sight to see hedge bottoms containing a mass of annual weeds, such as *Bromus sterilis* and *Galium aparine*. However, other factors may have increased the prevalence of species-poor field boundaries, such as disturbance from cultivations, spray drift and fertiliser contamination.

Weed distributions

Studies of the literature have indicated few reports of weed spread from field margins. The increase of a hedgerow species, *Bromus sterilis*, as a field weed in the United Kingdom is one well-documented (Froud-Williams, Pollard and Richardson, 1980). The spread of wind-borne seeds from land adjacent to fields, particularly road verges, has also been reported (Chancellor, 1969).

Studies of the plant distributions have been made on three arable farms in England. These are the Boxworth Experimental Husbandry Farm (EHF), Cambridge, the Manydown Estate, Basingstoke, and Bovingdon Hall, Essex. All three are on alkaline soils; the Boxworth EHF and Bovingdon Hall are on clay soils, while Manydown is on lighter chalky land. The mean number of plant species found in the boundary, the crop and common to both situations is given in Table 14.1.

Table 14.1   Mean number of plant species in the hedge, the field and in both locations on three farms

Number of fields sampled	Mean number of species		
	In hedge	In crop	In both hedge and crop
55	23.4	11.2	4.3

On average only a small proportion of the species found in the boundary also appear in the crop flora.

Four distribution patterns have been recorded for plant species at the edges of arable fields. These are as follows:

Type I: limited to the boundary, e.g. *Arum, maculatum, Dactylis, glomerata*;

Type II: usually in the crop, though occasionally in the boundary, e.g. *Veronica persica, Polygonum aviculare*;

Type III: in the boundary and at decreasing density into the crop, e.g. *Galium aparine, Bromus sterilis*;

Type IV: in crop and boundary with highest densities in the crop edge, e.g. *Alopecurus myosuroides*.

Those species with Type III patterns are those which may contribute to field weed populations and are of economic importance. However, many species only survive within 2.5 m of the boundary. Of the true field weeds which are adapted to disturbed ground (Type II), some also appear in the field boundary, taking advantage

Table 14.2 Number of sampled fields where common weeds were recorded, with the percentage of records where the species was found in the hedge only, in both hedge and crop or in crop only

Species	No. fields recorded	Percentage records in:		
		Hedge	Hedge+Crop	Crop
Elymus repens	52	50	42	8
Avena fatua	22	32	50	18
Alopecurus myosuroides	45	9	58	33
Poa trivialis	51	65	35	0
Bromus sterilis	44	59	36	5
Galium aparine	51	35	59	6
Viola arvensis	26	8	4	88
Convolvulus arvensis	42	55	31	14
Myosotis arvensis	29	41	25	34
Stellaria media	29	14	20	66
Polygonum aviculare	36	14	17	69
Fallopia convolvulus	34	3	26	71
Rumex obtusifolius	17	82	6	12
Cirsium arvense	37	76	19	5

of soil disturbance. These occurrences are included in the numbers of species common to crop and boundary given in Table 14.1. This further illustrates the low number of species with which the farmer may legitimately be concerned as posing a threat.

These studies on plant distributions have indicated the likely source, either boundary or within the crop, for the commonest weed species. After examining a 50m length of hedgerow and the adjacent crop area out to 50m on the three study farms, the locations of common weeds can be tabulated (See Table 14.2). Undoubtedly certain common grass weed species are capable of spreading from field edges, though most dicotyledonous weeds are maintained in the crop area. The fact remains that the bulk of boundary species do not appear in the crop and are therefore inoffensive.

Dispersal mechanisms

Plants are dispersed by a variety of means, though the reproductive adaptations are those of seed dispersal or vegetative propagation. In general the perennial species rely on vegetative propagation with some reliance on seed production followed by the maintenance of captured space by competition. Annual species rely on seeds surviving from one season to the next. Further success may be gained by seed dormancy or by biennial growth habit.

Vegetative spread is typically by rhizome growth, e.g. *Elymus repens*, or by the spread of viable root fragments or bulbs, e.g. *Allium vineale*. Seed dispersal can be passive, by wind, water, animals and man and his machines, or more rarely by active dehiscence, e.g. *Cardamine hirsuta*. The spread of *Elymus repens* (couch grass) from field boundaries was undoubtedly a problem to arable farmers in past times, especially as the rhizomes can grow a metre in a season (Palmer, 1958). However, the advent of pre-harvest treatments of glyphosate has rendered this spread academic. Most dispersal from field margins is by passive movement of seed usually only over short distances.

Implications for management

The present information indicates that only a small proportion of the boundary flora are significant weeds. Also, it has yet to be shown how significant are edge populations in contributing to or maintaining field populations. Work on seed dispersal by Fogelfors (1985) and Hume and Archibold (1986) indicate only limited seed movement from field edges. Given that only a small proportion of the flora are possible weeds, then management should recognise that most species are inoffensive and that, at most, selective control of weed species should be practised.

The common weeds of arable land that also occur in and spread from field boundaries are annual species. It makes ecological sense to encourage a perennial herbaceous flora, thus limiting the opportunity for such annual species.

Although not rigorously tested scientifically, it is generally accepted that increase in nutrient status and disturbance by close cultivation, spray drift and possibly the accumulation of trimmings after flail-cutting, have promoted a species-poor flora in many field boundaries. The dominance of the ground flora by a small number of species will often result from a high nutrient status. Taking these observations together, it would appear that management of the field boundary, so far as plants are concerned, should endeavour to reduce inputs and disturbance from adjacent agricultural operations, particularly fertiliser and pesticide contamination. The aim should be to develop a stable, species-rich habitat, which by its nature will contain only low populations of potential field weeds.

Recent initiatives in this area have been made by the Game Conservancy Cereals and Gamebirds Research Project, in which the outside 6 m of crop is not sprayed with most pesticides (Rands, 1985). This practice is currently being encouraged in West Germany and experiments are in hand in Denmark. Individual farmers in the United Kingdom are also experimenting with a range of barrier strips, including bare ground created chemically or mechanically. A programme to study different barrier strips is to be set up by the UK Ministry of Agriculture, Fisheries and Food. A grass strip between the crop and the boundary has been sown in by a number of farmers. The management of such strips is the subject of investigation at Long Ashton. Often these practices have been introduced where the boundary has serious infestations of *Bromus sterilis* or *Galium aparine*.

Further research

Several lines of research would increase our knowledge of the influence of the field boundary in agricultural land. Likewise, the effects of agricultural operations on adjacent uncropped areas are poorly understood. Few investigations have been made in grassland areas, where increased stocking rates and fertiliser use are probably affecting boundaries. Studies of the direct and indirect environmental effects of different pesticide applications and programmes in arable crops could usefully be made. For example, the knowledge of susceptibility of boundary species to herbicides is incomplete (Marshall, 1986). There may be opportunities for selective manipulation of weed species or boundary compositions with herbicides or plant growth regulators. The effect of fertilisers on boundary floras should be investigated and the relative importance of the different disturbance factors established. A broad-scale study of field boundaries on farms which have a range of pest control programmes, including organic farms, should also reveal interesting data.

Ecological investigations of dispersal of plants into agricultural land and along boundaries could lead to useful insights on the role of immigrations and emigrations on population processes. Studies on the contribution of boundary floras to field populations, including buried seed banks, require elaboration. The amounts of semi-natural habitat needed to maintain particular populations are unknown. This might be approached through studies of field size and hedge density. Techniques of managing field boundaries and boundary strips need refinement, and there is a particular need to devise methods of rehabilitating degraded, weed-infested field boundaries, perhaps including seed introductions.

References

Chancellor, R.J. 1969. Road verges – the agricultural significance of weeds and wild plants. In: *Road Verges, Their Function and Management*. Symposium. Monks Wood Experimental Station, Nature Conservancy Council, London. pp.29–35.

Fogelfors, H. 1985. The importance of the field edge as a spreader of seed-propagated weeds. *Weeds and Weed Control. 26th Swedish Weed Conference, 1985*, 178–89.

Froud-Williams, R.J., Pollard. F. and Richardson, W.G. 1980. Barren brome: a threat to winter cereals? *Report of the Weed Research Organization 1978–79*, pp.43–51.

Greaves. M.P. and Marshall, E.J.P. 1987. Field margins: definitions and statistics. In: *Field Margins*, eds, Way, J.M. and Greig-Smith, P.W. British Crop Protection Council Monograph, No. 35, pp. 23–34.

Hume, L. and Archibold, O.W. 1986. The influence of a weedy habitat on the seed bank of an adjacent cultivated field. *Canadian Journal of Botany, 64*, 1879–83.

Marshall, E.J.P. 1986. Studies of the flora in arable field margins. *Technical Report Long Ashton Research Station, Weed Research Division*, 96, 33 pp.

Marshall, E.J.P. and Smith, B.D. 1987. Field margin flora and fauna: interaction with agriculture. In: *Field Margins*, eds, Way, J.M. and Greig-Smith, P.W. British Crop Protection Council Monograph, No. 35, pp.23–4.

Palmer, J.H. 1958. Studies in the behaviour of the rhizome of *Agropyron repens* (L.) Beauv. 1. The seasonal development and growth of the parent plant and rhizome. *New Phytologist*, 57, 145–59.

Rands, M.R.W. 1985. Pesticide use on cereals and the survival of grey partridge chicks: a field experiment. *Journal of Applied Ecology, 22*. pp 49–54.

15 The role of field boundaries as reservoirs of beneficial insects

S.D. Wratten, Southampton University

The role and relative importance in pest suppression of the 300 or more species of polyphagous predators which occupy cereals in summer in the United Kingdom has been studied at Southampton, the Game Conservancy and elsewhere in recent years. There have been five aims to the work:

(a) Do polyphagous predators as a group contribute to pest suppression?
(b) Which species have the greatest ecological potential for pest suppression?
(c) Where and at what densities do these species overwinter on cultivated farmland and in 'natural' farmland habitats?
(d) What are the microclimatic and biotic requirements of the main overwintering species?
(e) (Can field boundaries be modified or partially re-created to enhance populations of beneficial insects, including polyphagous (and prey-specific) natural enemies?

The extent to which these questions have begun to be answered can be summarised as follows:

(a) Do they affect pest control? The following approaches have addressed this question and shown that the group does have such a role: simulation modelling (Carter and Sotherton 1983; Griffiths, 1983); field manipulation of their numbers (Edwards, Sunderland and George, 1979; Wratten and Pearson, 1982 (see Figure 15.1); Chiverton, 1986); diet analysis in relation to prey and predator density (Sopp and Chiverton, 1987).
(b) Which species are important? A battery of laboratory and field criteria have been combined to identify these: consumption rate

(Wratten, Bryan, Coombes and Sopp, 1984; Sopp and Wratten, 1986); phenology (Coombes and Sotherton, 1986 (see Figure 15.2); Coombes, 1986; Wratten et al, 1984); responses to prey aggregations —field (Bryan and Wratten, 1984); laboratory — video work (Wratten et al, 1984); climbing behaviour (Griffiths, Wratten and Vickerman, 1985); analysis of diet using enzyme-linked immunosorbant assay (Sopp and Chiverton, 1987). As a result of these studies, it is possible to identify species and groups of polyphagous predators which are of high biocontrol potential, at least as far as southern English cereals is concerned. These are: Staphylinidae (rove beetles): Tachyporus hypnorum; Carabidae (ground beetles): Agonum dorsale, Demetrias atricapillus, Bembidion lampros and B. Obtusum; Araneae (spiders): Family Linyphiidae, especially such genera as Erigone and Lepthyphantes; Hymenoptera-Parasitica, especially Aphidius spp.; Coccinellidae (lady-birds), especially Coccinella 7-punctata; Syrphidae (hoverflies), especially Episyrphus balteatus.

(c) Where do they overwinter? Sotherton's work at the Game Conservancy in the United Kingdom has shown the importance of field boundaries for many of the polyphagous Coleoptera species and which boundary types are most important (Sotherton, 1984; 1985). Carabidae, Staphylinidae and spiders dominate the fauna and can reach densities of 100 m^2 in winter in some boundaries. Many of the other arthropod groups mentioned above also depend to differing extents on the field boundary as an overwintering refuge or as a source of pollen, nectar or non-pest prey in spring and summer.

(d) Which aspects of the boundary are important for the predators? Many features of a boundary (physical, biotic, microclimatic) are irrelevant or even inimical to their role as invertebrate natural enemy reservoirs. While important features (largely sheltered, dry, micro-habitats) have been identified, the biological properties (such as age, density and species of herbaceous plants) of these favourable sites are unknown, together with the exact use made of them by over wintering predators.

(e) Can existing natural refuges be improved or can new refuges be created? The high densities of polyphagous predators which occur in some field boundaries (see (c) above) are often associated with mats or clumps of a range of grass species and the most favoured sites are often on raised earth banks typical of many hedge bottoms. For this reason, the woody components which usually define a hedge are not necessary for most polyphagous predators in winter. It is therefore possible to create the overwintering

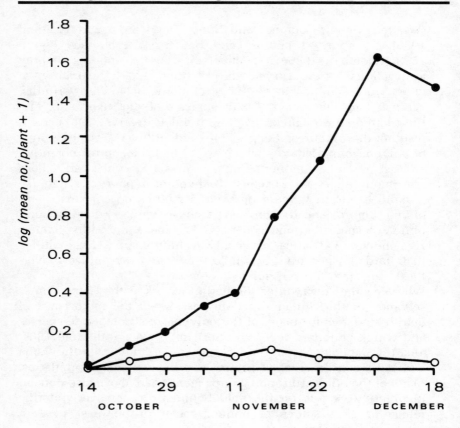

●predator numbers reduced by 50 per cent;
○control

Source: Wratten and Pearson, 1982.

Figure 15.1 Consequences of experimental reduction in predators' numbers for populations of the aphid *Myzus persicas* on New Zealand sugar beet.

habitats on farmland which favour the development of high populations of predators. The accumulation of such predators in the autumn and winter in these new 'island' habitats can then be monitored, together with their dispersal, distribution and predation rate in the crop in the spring and summer.

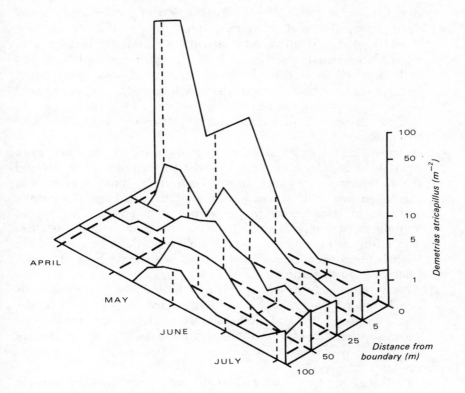

Figure 15.2 Density of the Carabid beetle (*Demetrias atricapillus*) at different distances from a field boundary; March–August 1984: suction samples
Source: Coombes and Sotherton, 1986.

Microclimatic conditions can be monitored in new and old, favourable and unfavourable boundaries, supported by laboratory work on microhabitat selection.

Current and proposed methodology

(1) Reduce field size experimentally by creating new, non-woody, predator overwintering refuges. Raised banks (created by careful ploughing) have been sown with grasses such as *Lolium, Dactylis,*

Agrostis and *Holcus*. Simple switching-off of the herbicide spray during normal cereal operations (plus the creation of adjacent sterile strips) will effectively provide new refuges hundreds of metres from existing orthodox field boundaries. Small (C.10 ha) and large (C.40 ha) fields have been used and 'peninsular' boundaries have been created which will reach the respective field centre. This type of bank can also be sown with pollen-bearing and nectar-bearing plants in order to attract Hymenoptera and Syriphidae.

(2) Increase the potential of existing boundaries. Preliminary work in 1986 has added refuges to existing boundaries, measured the microclimatic change brought about in those refuges and correlated this with overwintering predator numbers. The simple addition of straw bales to the sterile strip adjacent to a field boundary can create favourable microclimate refuges, for instance. The addition of a raised bank with grass, described above, or slightly-elevated boundary strip sown with grass (see Marshall, Chapter 14) can improve a hedge.

(3) Assess the effects of newly planted boundaries. Although this is a long-term study, such boundaries have been planted by the Game Conservancy and potential accumulation of predators is being studied already. At these and other study sites, boundary history is well known and will be used to try to interpret predator numbers.

In all the above cases, microclimatic and biotic changes brought about by the modifications made will be recorded. These will include recording the development of vegetative composition and cover; recording the microclimate in new or modified boundaries; recording the use of new or modified boundaries made by predators and the effects on their subsequent dispersal rate and distance and their predation rate in the crop.

In the current debate concerning food surpluses, in which ideas for reducing farming productivity or taking land out of production are being discussed, the ideas outlined above are highly relevant. Even if new 'boundaries' are created, as described, these represent a very small proportion of the cultivated area, require no planting of woody species, are easily created and easily maintained with selective herbicides. Each particular raised bank does not have to be a permanent feature; the direction of ploughing and drilling is changed on many farms, to minimise soil compaction, at four-to-five-year intervals. New raised banks can be created under such circumstances.

The cost of establishing the ridges across fields as described above, comprises three main aspects. The first is the labour involved in their physical establishment; the second is the establishment of the grass sward. These two actions together require 1–2 man-days. The third cost, that of the gross lost grain yield due to lost land, amounts to approximately 0.3 tonnes for a 10 ha field, worth some £30 in 1987. The benefits could arise from suppression of pest levels below economic thresholds; work at Southampton in 1987 has shown that a 50 per cent reduction in polyphagous predators numbers resulted in the grain aphids (*Sitobion*) numbers rising above the ADAS spraying threshold (J. Maurymootoo, personal communication). On the scale of a 10 ha field, this could avoid either an aphid-induced yield loss amounting to £300 (5 per cent yield loss at 6 t/ha and £100/t) if the crop is not sprayed with insecticide. Alternatively spray costs could amount to £150 for the field, costing labour and pesticide together. Also a ridge, once established, may persist for five years before ploughing direction is changed, so contributing to pest suppression over this period for one initial financial outlay.

References

Bryan, K. and Wratten, S.D. 1984. The responses of polyphagous predators to prey spatial heterogeneity: aggregation by carabid and staphylinid beetles to their cereal aphid prey. *Ecological Entomology*, 9, 251–9.

Carter, N. and Sotherton, N.W. 1983. The role of polyphagous predators in the control of cereal aphids. *10th International Congress of Plant Protection*, 778.

Chiverton, P.A. 1986. Predator density manipulation and its effects on populations of *Rhopalosiphum padi* (Hom: Aphididae) in spring barley. *Annals of Applied Biology*, 109, 40-60.

Coombes, D.S. 1986. The predatory potential of polyphagous predators in cereals in relation to timing of dispersal and aphid feeding. In *Ecology of Aphidophaga*, Hodek, I. (ed), 429–34. Academia, Prague and W. Junk, Dordrecht.

Coombes, D.S. and Sotherton, N.W. 1986. The dispersal and distribution of polyphagous predatory Coleoptera in cereals. *Annals of Applied Biology*, 108, 461–74.

Edwards, C.A., Sunderland, K.D. and George, K.S. 1979. Studies on polyphagous predators of cereal aphids. *Journal of Applied Ecology*, 16, 811–23.

Griffiths, E. 1982. The carabid *Agonum dorsale* as a predator in cereals. *Annals of Applied Biology*, 101, 152–4.

Griffiths, E. 1983. The feeding ecology of the carabid beetle *Agonum dorsale* in cereal crops. Doctoral dissertation, University of Southampton.

Griffiths, E., Wratten, S.D. and Vickerman, G.P. 1985. Foraging by the carabid *Agonum dorsale* in the field. *Ecological Entomology*, *10*, 181–89.

Sopp, P. and Wratten, S.D. 1986. Rates of consumption of cereal aphids by some polyphagous predators in the laboratory. *Entomologia Experimentalis et Applicata*, *41*, 60–73.

Sopp, P.I. and Chiverton, P. 1987. Autumn predation of cereal aphids by polyphagous predators in Southern England: a 'first look' using ELISA. *International Organisation for Biological Control/ W.PR.S. Bulletin* (in press).

Sotherton, N.W. 1984. The distribution and abundance of predatory arthropods overwintering on farmland. *Annals of Applied Biology*, *105*, 423–9.

Sotherton, N.W. 1985. The distribution and abundance of predatory Coleoptera overwintering in field boundaries. *Annals of Applied Biology*, *106*, 17–21.

Wratten, S.D. and Pearson, Joan 1982. Predation of sugar beet aphids in New Zealand. *Annals of Applied Biology*, *101*, 178–81.

Wratten, S.D., Bryan, K., Coombes, D. and Sopp, P. 1984. Evaluation of polyphagous predators of aphids in arable crops. *Proceedings British Crop Protection Conference*, 271–6.

Management of Grassland

Introduction

J.M. Way, Ministry of Agriculture Fisheries and Food, London

Grassland in all their forms are one of the major biotopes of the world, and are of great agricultural importance. They are also the habitats of very many species of wild plants and animals. In some parts of the European Community the conservation of these naturally occurring species is causing concern as a consequence of the intensification of use of grasslands for agricultural production. Figures presented at the Workshop for the extent of grasslands in a number of EC countries are summarised below:

Belgium: Agricultural land occupies more than 50 per cent of the national area, and 46 per cent of agricultural land is in grassland. Figures are given for use of grassland in the major farming areas showing wide variations throughout the country.

Denmark: Of 600,000 ha in grassland, 220,000 are in permanent grass and 360,000 in 1–2 year leys in an arable rotation.

Germany: 38 per cent of agricultural land is in permanent grass, but only about 3 per cent of this is extensive grassland of conservation interest. However, it is expected that about one million ha of grassland will come out of production in the foreseeable future, and it is intended that a proportion of this should be managed for environmental purposes.

Ireland: About 62 per cent of the national area is in agricultural grassland with an additional 14 per cent in rough grazing. Livestock units increased from 4.4 to 6.4 millions in the period 1968–80, as a consequence of an increase in grassland at the expense of arable land, and of the intensity of use of grassland.

Luxembourg: The area of permanent grass as a proportion of agricultural land has increased from 17 per cent in 1900 to 56 per cent in 1986, with arable land (including leys) declining from 80 per cent to 43 per cent over the same period.

Netherlands: About 70 per cent of the national area is farmed and more than 50 per cent is grassland, for which dry matter production/ha/year increased from 6,800 kg in c. 1950 to 11,000 kg in 1980, representing a considerable intensification of grassland use.

Great Britain: Agricultural land occupies 76 per cent of the national area, and over 60 per cent of the agricultural area is in grassland. There are strong variations in the proportions of grasslands in the major regions of the country. In England and Wales, figures suggest that there was a decline of 39 per cent of grassland area between 1932 and 1984, of which the unimproved or rough grasslands of greater significance for wildlife declined by 92 per cent. Figures available from the Nature Conservancy Council record very considerable losses of base rich, neutral (the Workshop was told during the visit to the Somerset Moors and Levels that there were only 750 ha of herb-rich lowland damp hay meadow remaining), and acid grasslands of wildlife interest.

The purpose of presenting these figures is to give some idea of the extent of agricultural grasslands in north-western Europe, of changes in the intensity of agricultural use, and an indication of the threat to permanent grasslands of conservation interest from intensification of agricultural use. They are also of interest in the context of objectives that might be sought for management of land where agricultural production may be reduced (as in Environmentally Sensitive Areas), or may cease (as in some Extensification Schemes), and to indicate the scope that exists for the enhancement of environmental interests. The fact that the figures are not presentable in a common format suggests that for many countries they are either not accessible or do not exist. In either case they will be required in the future if the environmental interest of the grassland resource is to be taken seriously, and will be particularly important to alert authorities where grassland types of high conservation value are at risk.

In the chapters in this part, descriptions are given by Maelfait of a Belgian study on the arthropod interest of marginal areas of grassland fields protected by a hedge or fence; by Schwaar of observations in Germany of the development of wildlife interest in areas of land ('fallow') that were no longer in agricultural use; and by Reyrink of the importance of meadow land in Holland for breeding birds, especially waders. In each chapter there are calls for protection of wildlife habitats on agricultural or previously agricultural land, and descriptions are given in the chapter by Wells and Sheail of the wider ecological importance of grasslands. Descriptions of the causes of loss of environmental interest are given in all the chapters either directly or as statistics (particularly by Hopkins and in several of the national reports) for temporal and spatial increases both in agricultural inputs and of production that are indicative of potential environmental damage.

As shifts occur in the Community Agricultural Policy, management systems of grasslands for agricultural production may reduce in intensity, so it will be important to identify objectives (agricultural production, wildlife conservation, landscape, recreation) for their future use and management. Management for environmental interests involves definition of objectives; for wildlife, objectives may differ between interests. At the workshop this was disputed in discussion, a fact that was itself interesting and suggests the need for continuing consideration amongst those whose concern is with the conservation of wildlife. Management of grasslands, and management costs, for environmental objectives are very much less well understood than for agricultural production, so that research is required if grasslands reverting from agricultural use are to be used to advantage for other purposes. This need is exposed in the chapter by Wells and Sheail. Consequently, information is required on the extent and potential of areas of grasslands in EC countries that are likely to be released from intensive agricultural production.

It is proposed that:

(i)   Exchange of research results between Member States, closer collaboration between research teams and further research are required on:
a. Agricultural Management of grasslands, for both agricultural production and environmental objectives.
b. Management of grasslands for environmental objectives where either land is no longer required for agricultural production, or it is desirable to recreate environmental interest, e.g. herb-rich chalk grassland.

(ii) Further research is required on autecology of individual species of plants and animals of particular interest in the context of managed grasslands.

(iii) Considerable areas of grassland will continue to be subject to intensive agriculture management, consequently research is required to identify opportunities that exist either in the fields themselves or at their margins (*vide* Maelfait *et al.*; and Reyrink) to conserve wildlife. Papers and discussions on the environmental, and notably wildlife, interest of agricultural grasslands in the EC should be a catalyst for further collaboration between experts; a Workshop would be a valuable way to progress item (i).

16

Utilisation of peatland for grassland and wetland habitat — contrast or agreement?

H. Kuntze and J. Schwaar, Niedersachsisches Landesamt fur Bodenforschung Bodentechnologisches Institut, Bremen

Permanent grassland

The advantages of permanent grassland are: yield capacity (up to 120 dt/ha), yield reliability (good water and nutrient dynamics), yield quality (raised bog grassland with lower incidence of parasites and toxic plants), and environmental advantages (low leaching of nutrients — water conservation). If grassland farming becomes too intensive, repeated cultivation leads to decreasing plant population density and loss of plant species diversity; together with risks of nutrient enrichment and a degrading effect on regional diversity, these have disadvantageous effects on wildlife and landscape. With a little exaggeration we might say that all EC grassland could become homogeneous throughout all the regions.

The advantages of grassland managed for a certain degree of naturalness (extensive grassland), are, in comparison, high plant population density and high plant species diversity, with the ability to adapt to extreme conditions (wetland habitat — dry habitat). Where peatland is used extensively as grassland it is an important biotope for exceptional species of plants and animals. Good soil and water conservation and climatic balance are associated abiotic effects.

Management

In order to preserve or restore a high diversity of wildlife and landscape interest, different methods of managing and ways of using natural landscapes in an harmonious manner are absolutely necessary. This is not only economically significant but ecologically necessary. This does not mean 'Back to Nature' at any price; but it means that we are ethically obliged to preserve and develop in an aesthetic way our man-made landscape which has developed over the last 6,000 years.

Nature conservation areas should not border closely on intensively used farmland. Here we see an opportunity for extensive grassland and fallow land to have a balancing and buffering effect. For example, the hydrological protection zone of peatland, which may alternate between 30–80 m (raised bog) and over 350 m (low moor with high timber content), provides such an opportunity.

If these transitional zones are to be used as farmland in the future, farmers will have to agree to a number of restrictions: no drainage, no cultivation or harvest between mid-May and end of June, reduced fertiliser usage, no chemical plant protection or ploughing up. This means financial losses up to 900 DM/ha depending on the relative level of output. In the Federal Republic of Germany, farmers get compensation up to 600 DM/ha for restrictions on utilisation.

Land with a certain degree of naturalness (extensive grassland, fallow land) has many environmentally beneficial effects. But some disadvantages should not be concealed. A run-off nearly twice as much as on intensively used grassland was measured on fen fallow land due to higher evapotranspiration. The extreme temperatures of the A-horizon of fallow land are always 1–3ºc lower than of the original grassland. According to model-calculations the wind velocity on rewetted peatlands without high vegetation rises with increasing area. This is found not only with treeless and scrubless natural habitats but also with cleared man-made landscapes.

Species distribution

The increase of species on former moor-grassland after many years of fallow can be estimated very positively. In a series of observations, an increase of up to thirty species within ten years was recorded. Subsequently, a decrease of species following a short-term increase of brackish marsh was due to aggressive competition from a few plant species. Woodland regeneration has only been observed from suckering near hedges and woods on wet fallow land, up till now. Floristic rarities (Red List) have not occurred even after a ten-year fallow period although potential sources of propagules are nearby. This can be correlated with continuing high nutrient status of the soil, causing a high production of biomass of common plants. The nutrient status is still higher than in natural plant associations. Threatened plant species (Red List) as well as some less threatened, but which had a certain degree of naturalness, were planted on the fallow land. Besides some failures such as *Arnica montana*, other species including *Serratula tinctoria*, *Scorzonera humilis*, and *Dianthus deltoides* have shown a tendency to spread. Other measures for the conservation of wildlife are also being tested. We know that such undertakings are controversial and looked on as adulteration of wildlife. But we should bear in mind that weak and reduced populations of plant species can only be preserved in this way.

Conclusion

Summarizing, it must be considered that (as shown by examples on peatland) disused farmland, extensively used grassland and intensive farming can be brought into line. In fact, all of these are necessary for a better and more beautiful environment. This opportunity should be exploited by economically induced extensification of grassland in the EC countries.

17

Bird protection in grassland in the Netherlands

L.A.F. Reyrink, Ministry of Agriculture and Fisheries, Utrecht

The Dutch meadow bird community

Meadow birds can be defined as bird species that breed predominantly in grassy vegetation such as pastures, meadows and hay fields. The grass-lands where the meadow bird community breeds, are all characterised by

Table 17.1 Breeding bird species in Dutch grasslands

I	Duck species	Mallard	*Anas platyrhynchos*
	Shoveler	*Anas clypeata*	
	Garganey	*Anas querquedula*	
	Tufted duck	*Aythya fuligula*	
II	Gallinaceous birds	Corncrake	*Crex crex*
	Partridge	*Perdix perdix*	
	Quail	*Coturnix coturnix*	
III	Waders	Lapwing	*Vanellus vanellus*
	Oystercatcher	*Haematopus ostralegus*	
	Black-tailed godwit	*Limosa limosa*	
	Curlew	*Numenius arquata*	
	Redshank	*Tringa totanus*	
	Ruff	*Philomachus pugnax*	
	Common snipe	*Gallinago gallinago*	
	Avocet	*Recurvirostra avosetta*	
IV	Songbirds	Skylark	*Alauda arvensis*
	Meadow pipit	*Anthus pratensis*	
	Yellow wagtail	*Motacilla flava*	
	Whinchat	*Saxicola rubhetra*	
	Stonechat	*Saxicola torquata*	
	Corn bunting	*Emberiza calandra*	

Source: Derived from Beintema, 1985

some kind of agricultural use. In fact the existence and maintenance of these grasslands and the community of meadow bird populations in the Netherlands depend on it. Four groups of breeding birds on grassland can be distinguished (see Table 17.1). The song birds breeding in grasslands are small and therefore inconspicuous. The three gallinaceous birds mentioned are very rare in the Netherlands. The ducks and waders on the other hand are very well known; especially the waders because of their size and characteristic behaviour. In the Netherlands when one is speaking of meadow birds it is mostly waders breeding in grassland that are meant. Although these birds also can be found breeding in other countries, the populations reached in the Dutch grasslands are unique (see Table 17.2). Accordingly the meadow bird community is a highly characteristic and esteemed element of the Dutch avifauna. From the total world population of black tailed godwits about 90 per cent breeds in the Netherlands. Therefore the Netherlands have an international responsibility for the conservation of the meadow birds.

Table 17.2 Estimated numbers of breeding pairs (x 1,000) of the six most important meadow birds in the Netherlands and their relative abundance in four European regions

	The Netherlands	Great Britain & Ireland	Scandinavia & Denmark	Belgium & France	Central Europe
Oyster-catcher	70	39	2.1	0.1	0.8
Lapwing	170	65	12	4.2	5.8
Black-tailed godwit	100	0.01	0.04	0.1	0.6
Redshank	27	5.8	3.2	0.04	0.9
Ruff	1.3	0.0	6.0	0.1	0.4
Common snipe	5.6	13	11	4.4	0.4

Source: derived from Beintema, 1985.

In Appendix 17.A the breeding distribution in the Netherlands of four meadow birds is shown. The meadow birds are distributed all over the country. The highest breeding desities are found in the wet and moist grasslands in the west and north-west.

At the beginning of this century there was a large increase in meadow bird populations. This was due to an increase in biomass of the soil fauna from the increasing use of fertilisers. However, over the last two decades agricultural intensification of dairy farming has meant that populations of meadow birds in the Netherlands have decreased strongly or have even become endangered. Because the international importance of the Netherlands for meadow birds has been widely recognised by the government and by nature conservation organisations, protection of meadow birds started at an early stage.

Most of the Dutch breeding meadow birds can be regarded as summer visitors since they migrate in autumn to winter south of the Netherlands, and return in the early spring. As an example, Appendix 17.B shows recovery sites of Black-tailed godwits and lapwings ringed as chicks in the Netherlands. The Black-tailed godwit winters south of the Sahel region in the estuaries of the rivers Senegal, Gambia and Niger (Speek and Speek, 1984).

In order to keep meadow bird populations in the Netherlands at a constant level, it is necessary that the mortality of adults is compensated by the number of juveniles. Because the majority of the meadow birds winter outside the Netherlands, we are not able to influence the mortality of adults. But because of an increasing interest in bird conservation in the wintering countries, an increase in mortality during migration and wintering is not very probable. Consequently, it appears that the decrease in breeding numbers of meadow birds is probably due to a decrease in nesting success, and a decrease in the survival rate of chicks. From yearly counts for some regions it can be shown that the ratio of juveniles to adults is decreasing (Biologische Stationen, 1983). It is considered that the decrease in the population of meadow birds is a result of the intensification of agricultural grassland use.

The amount of suitable breeding habitat is decreasing as a result of improving drainage and high usage of fertiliser. This leads to an earlier grass growth in spring. Therefore cattle will be grazed earlier and mowing will take place earlier. For meadow birds this means an increase in loss of eggs and chicks from trampling and machinery, and the period without agricultural use during the breeding season becomes too short for meadow bird species to produce enough offspring.

The various meadow bird species show a different timing in breeding. Species that breed early in spring are less vulnerable to intensification of agricultural use than species that start breeding later. Among the various meadow birds there is also a difference in the length of the breeding period. Figure 17.1 gives the hypothetical range of tolerance of some meadow birds to agricultural intensity.

Since the damaging effects of agricultural intensification became clear, the policy on meadow bird protection in the Netherlands has been to designate a number of strictly controlled reserves. Government and non-government nature conservancy organizations have bought farmland for this purpose and management has been either performed by the nature conservancy organizations themselves, or by farmers for a low rent.

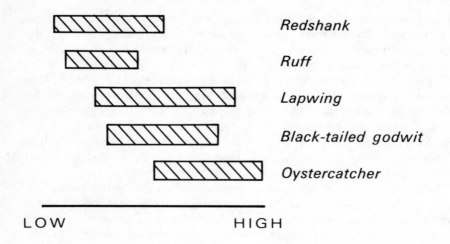

Figure 17.1   Hypothetical range of tolerance of some meadow birds to agricultural intensity

Source: derived from Beintema, 1985.

The policy of the 'Relationship' memorandum introduced in 1975 was to designate 50,000 hectares as nature reserves (by purchasing the farmland) and a further 50,000 hectares as management areas in which farmers could voluntarily conclude management agreements with the Government (De Boer and Reyrink, 1987). In the designated nature reserves farmers could also conclude management agreements until the time that they wanted to sell the land. For the restrictions on their farming operations arising from the management agreements the farmers were to receive financial compensation. The drafting of the first management schemes for these areas was attended by much discussion on the restrictions necessary for meadow bird conservation. To gather more information on possible and effective restrictions, the Dutch Bureau for Land Management in co-operation with the Research Institute for Nature Management started a research programme on meadow birds.

Research on meadow bird management

The first item to be researched was the influence of grazing cattle on breeding success. On grasslands with various densities of grazing cattle, survival of nests of meadow birds was measured. After finding a bird's nest it was visited once a week during the breeding season until the chicks had left the nesting site. With every visit the number of eggs in the nest and their condition was checked. At the same time all the agricultural activities were recorded, including length of periods of grazing. Table 17.3 gives a summary of the trampling of birds nests by grazing cattle. If the grazing density was more than twenty cows or yearlings to a hectare nearly all nests were trampled within a few days. Yearlings damaged more nests than cows and sheep, probably because of their wilder behaviour. There were also differences between the meadow bird species. Redshank and Black-tailed godwit were much more vulnerable to trampling by grazing cattle than oystercatcher and lapwing. This was probably due to the preference of the first two species to nest in long grass. Sites with long grass will be grazed relatively longer and more frequently. Another reason can be the more effective way of defending the nesting site by oystercatcher and lapwing against cattle.

From the results of this research we concluded that in management agreements for meadow bird conservation grazing with cattle should not be allowed during the breeding season. In situations where,

Table 17.3 Percentage of bird's nests trampled by grazing cattle with various densities and grazing periods

| | | Number of animals/ha | | | | | | | | |
| | | cows | | | yearlings | | | sheep | | |
Grazing period	Nesting birds	5	10	20	5	10	20	5	10	20
4 days	Oystercatcher	—	41	65	—	32	54	—	5	9
	Lapwing	—	31	53	—	57	81	—	14	26
	Black-tailed godwit	—	48	72	—	84	97	—	23	41
	Redshank	—	67	89	—	84	98	—	23	41
1 week	Oystercatcher	37	61	84	29	49	74	4	8	15
	Lapwing	28	48	73	52	77	95	12	23	40
	Black-tailed godwit	43	68	90	80	96	100	21	37	60
	Redshank	62	85	98	80	96	100	21	37	60
2 weeks	Oystercatcher	61	84	—	49	74	—	8	15	29
	Lapwing	48	73	—	77	95	—	23	40	65
	Black-tailed godwit	68	90	—	96	100	—	37	60	84
	Redshank	85	98	—	96	100	—	37	60	84

because of the microrelief, fields cannot be mown, but are always grazed, animal densities should be less than two cows or yearlings or six sheep, per hectare.

As part of an interdisciplinary study on the integration of nature conservation and agriculture, meadow bird research in the Netherlands was carried out. In this work the emphasis was on the length of the breeding period, hatching dates and the laying of replacement clutches after loss of the original nests. Information on hatching and breeding was necessary in order to know at what time farmers could graze cattle or mow the grass safely without damaging nests of meadow birds. The method was nearly the same as during the trampling research. In addition, during the course of searching for nests and checking them every week the incubation stages of the eggs were determined by means of an incubometer. This method is based on the loss of weight during incubation due to evaporation (Van Paassen *et al.*, 1984).

In Figure 17.2 the hatching periods of five meadow bird species are shown. For the common snipe no data are given because too few nests were found. From Figure 17.2 we can learn that in the research area 90 per cent of the lapwing and Black-tailed godwit clutches hatched in the first week of June. Of the oystercatcher and ruff, 90 per cent of the clutches hatched respectively around the 10 and 20 June.

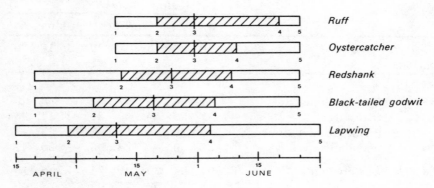

Figure 17.2 Hatching dates of some meadow bird species

1	hatching date of the first nest	4	90 per cent of all nests hatched
2	10 per cent of all nests hatched	5	hatching date of the last nest
3	50 per cent of all nests hatched		

For meadow bird management these results indicate that in the case of the presence of ruffs, mowing or grazing should not take place before the 20 June. In the case of lapwing and Black-tailed godwit, mowing dates based on hatching dates can be some time earlier, around the beginning of June.

Management agreements for meadow bird conservation therefore contain the following restrictions:

—no grazing or mowing before 1, 8, 15 or 22 June or 1 July depending on the species that are present;
—lowering of the ground water-table is not allowed;
—no breaking up of the grassland vegetation;
—during the breeding season no use of slurry, no harrowing or rolling.

The influence of agricultural use on breeding success on Dutch grassland is quite clear. Because we do not know the effects of mowing and grazing of grassland in which meadow bird chicks are present we cannot yet judge the effects of management agreements on population levels.

To answer this last question we are now concentrating research on the chick stage. By ringing (with coloured rings) we are measuring the survival rate of chicks. As long as the percentage of chicks that survive is unclear, we cannot predict whether the restrictions in the management agreements are effective from the point of view of meadow bird management.

Besides this ecological population research on meadow birds, the Bureau for Land Management in some regions has started an evaluation of management agreements. In these regions the number of breeding meadow birds is regularly counted in order to find out whether the desired population levels pursued are being reached. Within a few years we will know whether meadow bird management within management agreement areas on dairy farmland is achieving the goal we are aiming at, which is the conservation of the meadow bird community in the Netherlands.

References

Beintema, A.J., 1985. The Dutch meadow bird community. *Research Institute for Nature Management. Annual Report 1985*; 130–6.

Biologische Stationen, Rieselfelder Munster und Zwill brock, 1983. Zur Bestandsentwicklung der Uferschnepfe (Limosa, limosa) in Westfalen. *Ber. Dtsch. Sekt. Int. Rat. Vogelschutz*, 23, 121–8.

Paassen. A.G. van. Veldman, D.H. and Beintema, A.J., 1984. A simple device for determination of incubation stages in eggs. *Wildfowl*, 35, 173–8.

Speek, B.J. and Speek, G., 1984. *Thieme's vogeltrekatlas*. BV W.J. Thieme and Cie, Zutphen.

Teixeira. R.M., 1979. *Atlas van de Nederlandse broedvogels*. Vereniging tot behoud van Natuurmonumenten in Nederland, 's-Graveland.

Appendix 17.A: Breeding distribution of four medow bird species in the Netherlands
(These distribution maps were derived from Teixera, 1979)

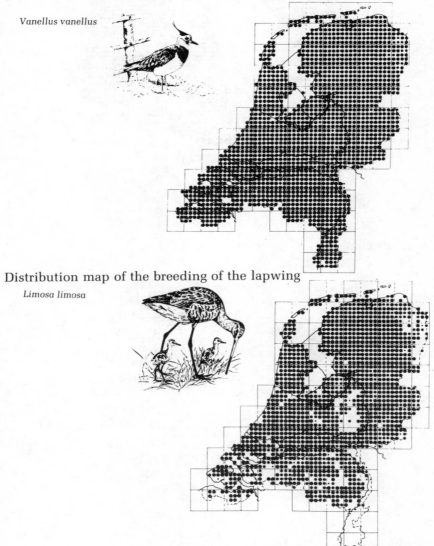

Vanellus vanellus

Distribution map of the breeding of the lapwing

Limosa limosa

Distribution map of the breeding of the Black-tailed godwit

Appendix 17.A: continued

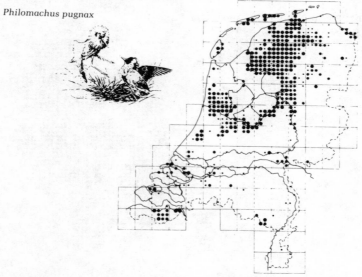

Tringa totanus

Distribution map of the breeding of the redshank

Philomachus pugnax

Distribution map of the breeding of the ruff

Appendix 17.B: Ring recoveries of Black-tailed godwit and lapwing
Source: derived from Speek and Speek, 1984.

Recoveries of Black-tailed godwit ringed as chicks in the Netherlands

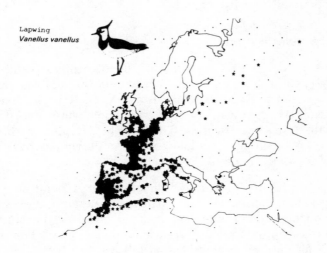

Recoveries of lapwing ringed as chicks in the Netherlands

18 | Investigations on the Arthropod fauna of grasslands

J.P. Maelfait, Instituut voor Natuurbehoud, Hasselt K. Desender, R. De Keer and M. Pollet, Laboratory for Ecology, Ghent.

The aim of this chapter is to give a survey of the investigations concerning Arthropods of grasslands that have been carried out by the Pedobiological Section of the Laboratory for Ecology, Zoogeography and Nature Conservation of the State University of Ghent. Results relevant to rural planning, recognizing the importance of the conservation and enrichment of the natural fauna and flora in a cultivated countryside, will be summarised and illustrated.

Survey of the investigations

In 1979 we started our investigations of the Arthropod fauna of an intensively grazed pasture, and of the border zone along its fence. The study plot is part of an Experimental Farm at Melle (Eastern Flanders, Belgium). From 1982 onwards other grasslands were incorporated (see Table 18.2) in our research. These were sampled within the framework of a landscape ecological study of rivulet valleys of Western and Eastern Flanders.

Our ultimate aim is to obtain a detailed understanding of the structure and functioning of the Arthropod communities of the grasslands concerned. For practical reasons however the study up till now has concentrated on the Araneae and Carabidae. The results reached thus far can be summarised as follows:

The address of the latter three contributors is: Laboratorium voor Ecologie, Zoögrafie en Natuurbehoud, Ledeganckstraat 35, 9000 Gent, Belgium.

(1) Comparative research on different sampling methods.

During the first years of investigation a lot of effort went into a large-scale and laborious empirical test on the use of fenced pitfall traps. Our results proved the method to be much better than the more widely used unfenced traps: fenced pitfalls give reliable relative density estimates for species occurring at low densities. Other sampling methods were also used: quadrat samples, time sorting pitfall, window traps, hand-collecting and suction sampling.

References: Desender et al., 1982a; Desender and Maelfait, 1983; Desender and Maelfait, 1986.

(2) Analysis of the structure of taxocoenoses.

References: De Keer et al., 1986a; Desender et al., 1981; Desender, 1982; Desender et al., 1982a; Desender et al., 1984b; Desender, 1985; Desender, 1986; Desender and Pollet, 1986; D'Hulster and Desender, 1982; D'Hulster and Desender, 1984; Maelfait et al., 1986a; Maelfait et al., 1986b; Maelfait and Segers, 1986; Maelfait et al., in prep.

(3) Analysis of the life cycle and population of the abundant spider and beetle species.

References: De Keer and Maelfait, 1986a; De Keer and Maelfait, 1986b; De Keer and Maelfait, in prep.,b; Desender, 1983; Desender and Crappe, 1983; Desender and Panne, 1983; Desender et al., 1984a; Desender and Pollet, 1985; Desender et al., 1985; Desender and Pollet, in press.

(4) Study of the feeding ecology of the abundant species.

Crop analysis of Carabid beetles, collection of the content of spider webs and hand-collecting of spiders with prey items between their chelicerae were used to assess the prey selection of these organisms. To evaluate the impact of this predation laboratory and field experiments were set up.

Reference: De Keer and Maelfait, in press; De Keer and Maelfait, in prep., a; Pollet and Desender, 1985; Pollet et al., 1985; Pollet and Desender, 1986; Pollet et al., 1986; Pollet and Desender, in press, a; Pollet and Desender, in press, b; Pollet et al., in press.

The arthropod fauna of an intensively grazed pasture and its margin

Our study plot of approximately 100×100 m is part of an intensively grazed pasture of 100×500 m. A drainage ditch runs along one of the long sides of the pasture. Between the ditch and the pasture and beneath a fence is a narrow zone (0.5 m width) which is only partly grazed, and is not trampled. The dominant grass of the pasture is perennial ryegrass (*Lolium perenne*) with a ground coverage of some 90 per cent. Dominant grasses of the edge are: *Festuca rubra, Holcus lanatus, Anthoxanthum odoratum* and *Festuca pratensis*; herbs that occur are: *Achillea millefolia, Anthriscus sylvestris, Heracleum spondylium, Rumex acetosa* and *Stellaria graminea*.

Table 18.1 Numbers of the indicated taxa present per pitfall trap (operative during a year) in the pasture and its margin

	Spider species	Spider families	Carabid species	Carabid families
Pasture	19	4	14	8
Margin	31	7	20	10

Table 18.2 Main features of six grasslands and number of species and families of spiders and number of species of Carabids per pitfall trap (operative during a year)
—   Grasslands

	A	B	C	D	E	F
Soil texture	clay	clay	clay	clay	loamy sand	sandy loam
Soil water content	high	high	high	high	low	low
Manuring	(+)	+	(+)	+	+	+
Haying	1×	2(3)×	—	1×	—	—
Cattle grazing	—	—	+	(+)	++	++
Diversity grasses	high	low	high	high	low	low
Diversity herbs	high	low	high	high	low	low
Litter layer	++	—	++	+	—	—
Juncus sp.	+	—	++	—	—	—
Carex sp.	++	(+)	+	—	—	—
Lolium perenne	(+)	++	(+)	+	++	++
Spider species	36	27	29	30	19	19
Spider families	7	4	5	4	4	4
Carabid species	23	27	21	26	16	14

+   Yes/present
++ in large amounts
(+) Limited
−   No/not present
1× once
2(3)× two or exceptionally three times

Table 18.3 Percentage abundance of the most frequently captured spider species in six grasslands*

Grasslands

	A	C	D	B	F	E
Alopecosa pulverulenta	9	59	0	0	5	27
Pardosa pullata	15	68	0	8	4	5
Diplocephalus permixtus	20	49	31	0	0	0
Oedothorax tuber-/gibb.	19	60	10	10	0	0
Lophomma punctatum	11	90	0	0	0	0
Pirata piraticus	23	62	0	15	0	0
Gnathonarium dentatum	42	57	0	0	0	1
Pachygnatha degeeri	3	33	29	18	13	4
Centromerita bicolor	6	9	67	8	4	6
Arctosa leopardus	0	100	0	0	0	0
Pardosa amentata	51	18	27	5	0	0
Dicymbium brevisetosum	37	14	46	0	0	3
Pirata latitans	51	38	0	6	0	6
Centomerus expertus	54	29	5	7	0	5
Ceratinella brevipes	67	33	0	0	0	0
Pachygnatha clercki	46	8	16	22	3	4
Oedothorax retusus	33	13	11	11	4	29
Agyneta decora	54	1	20	14	3	0
Walckenaera nudipalpis	39	4	56	0	1	0
Gongylidiellum vivum	32	34	9	23	1	0
Leptorhoptrum robustus	34	18	0	7	0	40
Centromerus sylvaticus	64	8	28	0	0	0
Pardosa palustris	0	1	70	4	4	21
Lepthyphantes tenuis	8	2	57	10	23	0
Erigone atra	0	3	8	27	27	35
Erigone vagans	1	1	0	34	40	24
Oedothorax apicatus	1	0	0	29	50	20
Erigone dentipalpis	1	1	3	12	32	52
Bathyphantes gracilis	5	7	15	15	39	19
Oedothorax fuscus	5	6	14	12	16	48
Lepthyphantes insignis	0	0	26	7	67	0

*Species and sampling sites in the order obtained by the ordination method explained in Maelfait and Sepers 1986.

The total density of adult spiders in the pasture during summer reaches 150 to 200 individuals per m_2, and for Carabids 30 to 40 per m_2. Both taxocoenoses are strongly dominated by a few species. These are ubiquists overwintering in the adult stage. They show a high developmental rate and reproduction during summer (two generations a year for the most abundant spider species).

A much richer fauna occurs in the border zone. This is illustrated by Table 18.1 in which the number of species and higher taxa of spiders and Carabid beetles caught per year per pitfall trap (median values) are shown. It is evident that there are large number of species limited to this zone that are not able to survive in the adjacent pasture.

Another important aspect of the border zone is that it gives shelter during winter to quite a large number of Carabids and spiders of the pasture. Maximum densities observed during the cold season were 250 individuals for spiders and 900 for Carabids. The well-developed sod and litter layer of the border zone offers a well-aerated and warm hibernation site for many invertebrates.

We have concluded that the grassy edge of pastures is an important element in the agricultural landscape, and that it substantially contributes to the overall biotic richness of the cultivated countryside.

Carabids and spiders of different grassland types

The Arthropod communities of ten grasslands situated in Flanders (Belgium) were sampled by means of pitfall traps. To give an idea of the results of these investigations, the main features of six of these sampling sites together with measures of the richness of the Carabid and Araneid taxocoenoses are listed in Table 18.2 and results obtained for spiders are given in Table 18.3. It follows that the intensity of agricultural management (manuring and haying or grazing) has a profound influence on the invertebrate fauna of grasslands. Intensively managed grasslands have poor animal communities composed of eurytopic, opportunistic species. The rich Arthropod faunas of grasslands with only a marginal importance for agriculture can only be preserved if their traditional management is continued. Drainage and/or high levels of fertiliser usage would destroy them, and so would no management at all. This suggests that these man-made grasslands with a rich fauna should be placed under the control of nature conservation authorities.

References

De Keer, R. and Maelfait, J.P. 1986a. Field and laboratory observations on the life history pattern of *Oedothorax fuscus* (Blackwall) (Linyphiidae, Araneida). *Annls Soc. r. zool. Belg.*, *116*, 90–1.

De Keer, R. and Maelfait, J.P. 1986b. Reproduction and development of some ubiquitous Linyphiid spiders (Linyphiidae, Araneida) under laboratory conditions. *Annls Soc. R. zool. Belg.*, 116, 91–2.

De Keer, R., Desender, K., D'Hulster, M. and Maelfait, J.P. 1986a. The importance of edges for the spider and beetle fauna of a pasture. *Annls Soc. r. zool. Belg.*, 116, 92–3.

De Keer, R. and Maelfait, J.P. in press. Laboratory observations on the development and reproduction of *Oedothorax fuscus* (Blackwall, 1834) (Araneida, Linyphiidae) under different conditions of temperature and food supply. *Rev. Ecol. Biol. Sol.*

De Keer, R. and Maelfait, J.P. in prep., a. Laboratory observations on the development and reproduction of *Erigona atra* (Blackwall, 1841) (Araneae, Linyphiidae).

De Keer, R. and Maelfait, J.P. in prep., b. Life history of *Oedothorax fuscus* (Blackwall, 1834) (Araneida, Linyphiidae) in a heavily grazed pasture.

Desender, K., Maelfait, J.P. D'Hulster, M. and Vanhercke, L. 1981. Ecological and faunal studies on Coleoptera in agricultural land. I. Seasonal occurrence of Carabidae in the grassy edge of a pasture. *Pedobiologia*, 22, 379–84.

Desender, K. 1982. Ecological and faunal studies on Coleoptera in agricultural land. II. Hibernation of Carabidae in agro-ecosystems. *Pedobiologia*, 23, 295–303.

Desender, K., D'Hulster M., Maelfait, J.P. and Vanhercke L. 1982a. Onderzoek uit het Laboratorium voor Oecologie der Dieren, Zoogeografie en Natuurbehoud. Oecologie van weide-arthropoda. *Biol. Jb. Dodonaea*, 50, 45–50.

Desender, K., Maelfait, J.P. and Vanhercke, L. 1982b. Variations qualitatives saisonnieres des Carabidae (Coleoptera) d'une prairie paturee a Melle (Frandre Orientale, Belgique), etudies a l'aide de differentes methodes d'echantionnage. *Biol. Jb. Dodonaea*, 50, 83–92.

Desender, K. 1983. Ecological data on *Clivina fossor* (Coleoptera, Carabidae) from a pasture ecosystem. 1. Adult and larval abundance, seasonal and diurnal activity. *Pedobiologia*, 25, 157–67.

Desender, K. and Crappe, D. 1983. Larval and adult morphology and biometry of two sibling species *Bembidion lampros* (Herbst) and *Bembidion Properans* Stephens (Coleoptera, Carabidae). *Biol. Jb. Dodonaea*, 51, 36–54.

Desender, K. and Maelfait, J.P. 1983. Population restoration by means of dispersal, studied for different carabid beetles (Coleoptera, Carabidae) in a pasture ecosystem. *New Trends in Soil Biology*, Proceedings of the VIII. Intl. Coll. on Soil Zoology. Louvain-la-Neuve (Belgium), 30 August – 2 September 1982. Lebrun, Ph., H.M. Andre,

A. De Medts, C. Gregoire-Wibo and G. Wauthy eds., Dieu-Brichart, Ottignies-Louvain-la-Neuve, pp. 541–50.

Desender, K. and Panne, V. 1983. The larvae of *Pterostichus strenuus* Panzer and *Pterostichus vernalis* Panzer (Coleoptera, Carabidae). *Annls Soc. r. zool. Belg.*, *113*, 139–54.

Desender, K. and Panne, V. 1983. The larvae of *Pterostichus strenuus* Panzer and *Pterostichus vernalis* Panzer (Coleoptera, Carabidae). *Annls Soc. r. zool. Belg.*, *113*, 139–54.

Desender, K., Mertens, J., D'Hulster, M. and Berbiers, P. 1984a. Diel activity patterns of Carabidae (Coleoptera), Staphylinidae (Coleoptera) and Collembola in a heavily grazed pasture. *Rev. Ecol. Biol. Sol*, *21*, 347–61.

Desender, K., Pollet, M. and Segers, H. 1984b. Carabid beetle distribution along humidity-gradients in rivulet-associated grasslands (Coleoptera, Carabidae). *Biol. Jb. Dodonaea*, *52*, 64–75.

Desender, K. 1985. Graslandbeheer en invertebraten. *Natuurreservaten*, *7*, 88–91.

Desender, K. and Pollet, M. 1985. Ecological data on *Clivina fossor* (Coleoptera, Carabidae) from a pasture ecosystem. II. Reproduction, biometry, biomass, wing polymorphism and feeding ecology. *Rev. Ecol. Biol. Sol*, *22*, 233–46.

Desender, K., Van Den Broeck, D. and Maelfait, J.P. 1985. Population biology and reproduction in *Pterostichus melanarius* ILL. (Coleoptera, Carabidae) from a heavily grazed pasture ecosystem. *Med. Fac. Landbouww. Rijksuniv. Gent*, *50*, 567–75.

Desender, K. 1986. On the relation between abundance and flight activity in Carabid beetles from a heavily grazed pasture. *Z. ang. Ent.*, *102*, 225–31.

Desender, K. and Pollet, M. 1986. Adult and larval abundance from Carabid beetles (Coleoptera, Carabidae) in a pasture under changing grazing management. *Med. Fac. Landbouww. Rifksuniv. Gent*, *51/3a*, 943–55.

Desender, K. and Maelfait, J.P. 1986. Pitfall trapping within enclosures: a method for estimating the relationship between the abundances of coexisting carabid species (Coleoptera, Carabidae). *Holarct. Ecol*, *9*, 245–50.

Desender, K. and Pollet, M. in press. Life cycle strategies in the most abundant ground beetles from a heavily grazed pasture ecosystem. *Med. Fac. Landbouww. Rijksuniv. Gent*.

D'Hulster, M. and Desender, K. 1982. Ecological and faunal studies on Coleoptera in agricultural land. III. Seasonal abundance and hibernation of Staphylinidae in the grassy edge of a pasture. *Pedobiologia 23*, 403–14.

D'Hulster, M. and Desender, KI. 1984. Ecological and faunal studies on Coleoptera in agricultural land. IV. Hibernation of Staphylinidae in agro-ecosystems. *Pedobiologia*, 26, 65–73.

Maelfait, J.P., Desender, K., De Keer, R. and Pollet, M. 1986a. Ecological investigations on the arthropod fauna of agro-ecosystems at Melle (Eastern Flanders, Belgium). *Annls Soc. r. zool. Belg.*, 116, 103–4.

Maelfait, J.P., Seys, J., De Keer, R., De Kimpe, A., Desender, K. and Pollet, M. 1986b. Relations between the agricultural management on epigeic arthropod fauna of grasslands. *Annls Soc. r. zool. Belg.*, 116, 105.

Maelfait, J.P. and Segers, R. 1986. Spider communities and agricultural management of meadow habitats. *Actas X Congr. Int. Arachnol. Jaca (Espana)*, 1986, I, 239–43.

Maelfait, J.P., De Keer, R. and Desender, K. in prep. The arthropod community of the edge of an intensively grazed pasture. *Second Int. IALE-seminar*, Munster.

Pollet, M. and Desender, K. 1985. Adult and larval feeding ecology in *Pterostichus melanarius* ILL. (Coleoptera, Carabidae). *Med. Fac. Landbouww. Rijksuniv, Gent*, 50, 581–94.

Pollet, M., Desender, K. and Maelfait, J.P. 1985. Onderzoek naar de voedselkeuze van loopkevers (Carabidae, Coleoptera) in Weideo-ecosystemen. *Bull. Annls Soc. r. belge ent*, 121, 494–7.

Pollet, M. and Desender, K. 1986. Prey selection in Carabid beetles (Coleoptera, Carabida): are dual activity patterns of predators and prey synchronized? *Med. Fac. Landbouww. Rijksuniv. Gent*, 51/3a, 957–71.

Pollet, M., Desender, K. and Maelfait, J.P. 1986. Aspects of the feeding ecology of *Pterostichus melanarius* (Coleoptera, Carabidae) in a heavily grazed pasture. *Annls Soc. r. zool. Belg.*, 116, 110–11.

Pollet, M. and Desender, K. in press, a. The consequences of different life histories in ground beetles for their feeding ecology and impact on other pasture arthropods. *Med. Fac. Landbouww. Rijksuniv. Gent.*

Pollet, M. and Desender, K. in press, b. Feeding ecology of grass-land-inhabiting carabid beetles (Carabidae, Coleoptera) in relation to the availability of some prey groups. *Acta Phytopath. Entomol. Hung.*

Pollet, M., Desender, K. and Van Kerckvoorde, M. in press. Prey selection in *Loricera pilicornis* (Coleoptera, Carabidae). *Acta Phytopath. Entomol. Hung.*

19 | Distribution and Management of Grassland in the United Kingdom

A. *Hopkins*, Institute for Grassland and Animal Production, Okehampton, Devon.

The vegetation cover of most of the United Kingdom under natural conditions would be forest, but today grassland occupies over 60 per cent of the agricultural land and dominates the landscape in many northern and western parts of the country. The extent of the grassland area and the way it is farmed have changed considerably during historic times, but the period since 1940 has seen the most profound changes. Increases in stocking rate and the use of artificial fertilisers, reseeding of old pastures, underdrainage of wetter land and the widespread replacement of hay by silage have contributed to a general intensification in grassland farming, resulting in marked changes in the composition and character of grassland, and in the agricultural landscape.

Location and types of grassland in the United Kingdom

Of approximately 12.1 million ha of enclosed agricultural land in the United Kingdom, 5.3 million ha supports crops, mainly wheat and barley, and 6.8 million ha is in grass. In addition there are over 6 million ha of rough grazing, of which at least a quarter consists predominantly of grass species (Ministry of Agriculture, Fisheries and Food, 1986.)

The relative proportions of grass and arable land in different areas of the United Kingdom are shown in Figure 19.1. In the drier

The author wishes to thank colleagues in the Institute for Grassland and Animal Production and in the Departments for Agriculture in Scotland and Northern Ireland, whom he consulted whilst preparing this chapter.

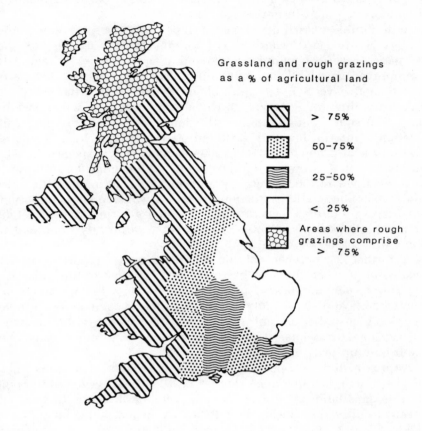

Grassland and rough grazings
as a % of agricultural land

> 75%

50-75%

25-50%

< 25%

Areas where rough
grazings comprise
> 75%

Figure 19.1 The division of agricultural land between tillage and grassland in the United Kingdom, 1986

Source: derived from MAFF June 1986 statistics.

lowland areas of east and south-east England over 80 per cent of the agricultural land is used for arable crops. Mixed arable and grassland farming is practised through central and southern England and parts of eastern Scotland. Much of the land which has been converted from grassland to arable in recent years is located in these areas. Grassland predominates in northern and south-west England and throughout much of Wales, Scotland and Northern Ireland. Within these areas most of the land above 600 m is occupied by rough grazings frequently composed of moorland species (e.g. *Calluna vulgaris, Vaccinium myrtilus*); similar vegetation occupies much of the unforested land in northern Scotland.

Agricultural grasslands range in type from recently sown and intensively managed swards of *Lolium* spp. through ageing leys and permanent swards containing sown and unsown species, to relatively unimproved grassland predominantly of unsown species. Swards of less than five years old occupy about 25 per cent of the grassland area (excluding rough grazings). Higher proportions of young swards occur in south and south-east England (often within an arable rotation), and in lowland Scotland and Northern Ireland. These swards usually contain at least 80 per cent sown species, mostly *Lolium perenne*. Sowings of short-term *Lolium multiflorum* swards have declined in recent years but are locally important as the grass phase in arable rotations. *Phleum pratense* is considered an important sown species in areas where winter hardiness is important; *Dactylis glomerata*, and more recently *Bromus* spp., are locally important in some drier eastern areas.

A further 25 per cent of the grassland area supports reseeded swards of between five and twenty years old. Most of this grassland is *Lolium perenne* based, often co-dominant with *Agrostis* spp., *Poa trivialis* and *Holcus lanatus* are often major components; *Rumex* spp. may also be present particularly under the more intensive management associated with dairying, and *Trifolium repens* may contribute up to 10 per cent of the sward.

Approximately 50 per cent of the grassland area, plus most of the rough grazing areas are at least twenty years old. Some of this grassland is of considerable age, particularly that which escaped cultivation during the 1939–45 wartime period. It includes the remaining areas of chalk downland, river meadows, coastal marshes and other wetlands, coastal headlands and parkland. Older swards in these situations may commonly contain ten or more grass species and a wide range of herb species. A high proportion of the United Kingdom's older grassland is in upland areas where physical problems and low economic returns have discouraged agricultural improvement. On many lowland farms there are areas of old grass-land which have remained because of physical difficulties or because owners have not been under economic pressure to carry out improve-ments. Nevertheless, a high proportion of the older grassland area has been agriculturally improved either through management alone or as a result of earlier reseeding. The presence of *Lolium perenne* is an indicator of agricultural improvement: swards of over twenty years of age in lowland Britain contain, on average, about 25 per cent of this species, and it is now absent from less than 10 per cent of lowland grassland of this age group. In the uplands most soils are inherently acidic and support *Festuca–Agrostis* swards on enclosed land on brown-earth soils or *Nardus stricta–Molinia caerulea* on peaty soils, with the more intensive management being confined to relatively small areas of 'in-bye' land adjacent to farmsteads. Floristically-rich hay meadows still occur on some of this land but much has been reseeded or otherwise improved. Grassland containing herb species is still relatively common in upland areas: a recent survey conducted by the author found over 30 per cent of upland grassland in England and Wales contained herb species, though relatively little of this could accurately be described as herb-rich.

Historical changes in the use, management and composition of grassland

The area occupied by grass has periodically declined with increased demands for arable crops: in the sixteenth and seventeenth centuries, in the first half of the nineteenth century and in the period since 1940. Conversely, the total grassland area has periodically increased when farming has been depressed, e.g. between 1874 and 1914 over 1.5 m ha of arable land returned to grass, in many cases without being sown.

Information on the botanical composition of grasslands, and on sward age and management, has been obtained from grassland surveys conducted at intervals since the 1930s. These refer specifically to the situation in England and Wales though might reasonably be extrapolated to other areas of the United Kingdom. In 1939 less than 10 per cent of the grassland supported sown 'temporary' swards. Of the remaining permanent grassland less than 6 per cent was described as first-grade or second-grade (e.e. *Lolium perenne* contributing 15 per cent or more of ground cover). *Agrostis–Lolium* pastures comprised a further 20 per cent of the area and *Agrostis* based pasture a further 55 per cent.

Ploughing of grassland during the war period reduced the permanent grassland area; subsequent reseeding to *Lolium perenne* based swards and more intensive management during the 1960s and 1970s have exerted further changes in botanical composition. Surveys on grassland farms conducted by the author in 1983 and 1986 have shown higher·average contributions of *Lolium perenne* in grassland than were reported from surveys conducted in the early 1970s (Green, 1982). At least 60 per cent of the grassland over twenty years old is now of comparable botanical composition to the first-grade and second-grade pastures of the 1930s — ten times the proportion recorded fifty years ago.

Present grassland management

Reseeding

Currently about 0.5 million ha of grassland are sown each year in the United Kingdom, i.e. about 8 per cent of the grassland area. Approximately 10–15 per cent of this area is sown to one-year leys, normally based on *Lolium multiflorum*. However, the major part is sown to *Lolium perenne* or mixtures based on this species. In recent years there has been a reduction in the area of grassland sown each year and an increase in the intended duration of sown swards. A recent (1986) survey of farms in upland areas of England and Wales found that of grassland which was known to have been over twenty years old in 1971, only 18 per cent had been reseeded since then. Farmers had, however, carried out further renewals of swards which had already been reseeded. Newly sown swards tend to produce higher yields during the first harvest year, but thereafter they have little yield advantage over most similarly managed older swards.

Grazing

Almost all agricultural grassland is grazed for at least some part of the year though there are wide variations in stocking rates. During the past fifty years the average stocking rates on grassland have almost doubled. United Kingdom grassland now provides about 80 per cent of the dietary requirements of about 18 million sheep plus their lambs, and 12.8 million cattle. Additionally there are an estimated 0.5 million horses, mainly kept for recreational purposes.

The intensity and seasonality of grazing, and the type of grazing livestock affect the composition of grassland. Intensive grazing is generally favourable to the development of *Lolium perenne* and contributes to the loss from the sward of most species of flowering plants.

Mowing

The seasonal growth of grass with peak production in early summer and little growth between October and March necessitates that a considerable area of grassland be mown for conserved forage. Approximately 40 per cent of the grassland area is mown every year or most years. Traditionally this has been as hay cut at a mature stage of growth from late June onwards but the widespread adoption of silage making in recent years has resulted in grass being cut much earlier in the season. In the early 1970s the relative ratios (by dry weight equivalent) of hay and silage were 85 per cent to 15 per cent. The corresponding proportions are now about 30 per cent to 70 per cent. Silage fields tend to be younger reseeded fields and to receive high levels of fertiliser nitrogen and other manures. This management, combined with the earlier mowing, produces swards composed predominantly of a small number of species such as *Lolium perenne*, *Poa trivialis* and perhaps *Holcus lanatus*, with most herb species being eliminated. Hay is still important in some areas particularly with livestock rearing in the upland areas.

Drainage

A large proportion of United Kingdom grassland is on soils derived from shales, clays or glacial drift which are inherently prone to seasonal waterlogging. There is a long history of land improvement by artificial drainage and during the nineteenth century about 5 million ha of land were under-drained. Many areas of wetland were converted to pasture or arable at this time. There was an active period of installation between 1940 and 1980, encouraged by grant aid, but the annual rate of field drainage may have fallen in the last few years (Baldock, 1984).

Improvements in drainage directly affect grassland through changes in the soil water regime leading to the demise of species associated with wet conditions. The sward is further affected by indirect changes in management, e.g. stocking rates increased and the grazing period extended, often with increased or earlier applications of fertiliser or the introduction of silage making. In mixed farming or arable areas grassland is often ploughed after drainage and converted to arable crops.

Herbicides and pesticides

Herbicides are widely used on grassland to check broad-leaved weeds in newly sown leys and to control perennial weeds in established swards, notably docks (*Rumex* spp.), thistle (*Cirsium* spp.) and buttercup (*Ranunculus* spp.). Approximately 40 per cent of grassland in England and Wales receives herbicide either as an all-over spray or more commonly as spot-treatments against localised weeds (Sly, 1986). When allowance is made for the area of spot-treatments the actual area treated is lower: one report suggests only 5 per cent was treated in 1986, a marked reduction over earlier years (Parrott, 1987).

On established grassland the use of fungicides and molluscicides is minimal and likely to remain so. Insecticides are mainly used to control leather jackets (larvae of *Tipula* spp.) and although in most years only about 20–30,000 ha are treated, up to 0.5 million ha may warrant treatment in an outbreak year. The use of pesticides on grassland is very low in comparison to that on cereal crops. Grassland farmers do not often see the need for agrochemicals and in most years the cost of the treatment is likely to be regarded as high in relation to the value of the crop. In the future the most likely increases are the adoption of insecticidal and fungicidal seed treatments for newly sown leys, now that cheap formulations are available.

Fertilisers

The increased use of inorganic fertilisers, particularly nitrogen, has been one of the most important trends associated with the intensification of grassland farming in recent years. The amount of fertiliser nitrogen applied to United Kingdom grassland has more than doubled in the past twenty years. Currently about 85 per cent of the grassland area receives some fertiliser nitrogen, the average rate applied being about 150 kgN/ha (Chalmers and Leech, 1986; Kershaw, 1986). Grassland mown for hay commonly receives less than 100 kgN per ha whilst that mown for silage or used for paddock-grazing or strip-grazing on dairy farms typically receives 250–350 kgN/ ha. On average older swards receive less fertiliser nitrogen than younger sown swards. Grassland which receives no fertiliser is often affected by features which restrict the use of farm machinery: slope, wetness, accessibility.

Approximately 70 per cent of grassland also receives phosphorous and potassium, usually in the form of compound fertilisers, and typically at rates which provide 20–40 kg/ ha of P_2O_5 and K_2O. Additional inputs of these elements are returned through dung and urine. Except on calcareous soils the pH is often below optimum for grassland production and periodic application of ground limestone to raise pH to about 6.0 is common practice. Each year about 5 per cent of the grassland area receives this treatment.

The application of high levels of fertiliser, particularly nitrogen, leads to a reduction in species diversity and usually results in an increase in the proportion of the sward contributed by *Lolium perenne*; *Poa trivialis* and *Holcus lanatus* may also increase, with most other grass species either reduced or eliminated. *Trifolium repens* is uncompetitive under high nitrogen but potassium and phosphorus fertilisers generally favour its development. Most of the flowering species associated with traditionally managed grassland are eliminated under management which includes high inputs of fertiliser.

Environmental Considerations

Whilst most grassland is managed primarily for agricultural production, many of the grassland areas of the United Kingdom are also of scenic and wildlife interest, including parts of National Parks and designated Areas of Outstanding Natural Beauty. Many aspects of modern grassland management are incompatible with landscape and wildlife conservation and there is growing public concern about the impact of modern farming. Legislative procedures have been introduced to protect designated areas, including Sites of Special Scientific Interest (SSSIs) administered by the Nature Conservancy Council against detrimental change and many other such areas are managed by local authorities and voluntary conservation bodies. In 1986 the United Kingdom government announced the setting up of Environmentally Sensitive Areas (ESAs) in which compensation arrangements are to be offered to farmers in return for pursuing traditional, environmentally – sympathetic methods of farming, especially of grasslands.

References

Baldock, D. 1984. *Wetland Drainage in Europe*. Institute for European Environmental Policy.

Chalmers, A.G. and Leech, P.K. 1986. *Fertilizer Use on Farm Crops in England and Wales, 1985*. (Rothamsted Experimental Station).

Green, J.O. 1982. *A Sample Survey of Grassland in England and Wales*. (Grassland Research Institute, Hurley).

Kershaw, C.D. 1986. *Report on the Survey of Fertilizer Practice, Scotland 1985*. (Rothamsted Experimental Station).

Ministry of Agriculture Fisheries and Food. 1986 (and previous years). Annual census returns.

Parrot, T. 1987. Weed control in Grassland. *Grass Farmer*, 26, 12–3.

Sly, J.M.A. 1986. Review of usage of pesticides. *Survey Report*, 41 (Harpenden Laboratory).

20 | The effects of agricultural change on the wildlife interest of lowland grasslands

T.C.E. Wells and *J. Sheail, NERC
Institute of Terrestrial Ecology,
Monks Wood, Huntingdon*

Recent research in archaeology and palaeoecology has drawn attention to the long-standing and extensive impact of early man on the natural environment. The natural woodland cover was destroyed or greatly modified, and wide areas cultivated. Soil erosion brought about large-scale changes, both on the higher ground and in those parts of the lower valley systems where the eroded material was deposited (Robinson and Lambrick, 1984). Throughout this time, the extent and character of grasslands have varied, according to the objectives and resources of farmers and other land users. It is, however, only in the last 300 years or so that there is significant evidence of farmers consciously trying to change the composition of swards through the use of different seeds-mixtures and through adjustments to grazing, mowing and burning practices.

Historic changes in the area of grasslands

It is far from easy to trace changes in the extent and character of grasslands. Because of the size and shape of the parish unit, the

Miss L. Farrell of the Chief Scientist's Directorate, Nature Conservancy Council, very kindly provided data on rare plant species. This chapter draws in part on research commissioned by the Nature Conservancy Council from the NERC Institute of Terrestrial Ecology. The assistance of the Ministry of Agriculture, Fisheries and Food in carrying out the research is gratefully acknowledged.

annual census of the Ministry of Agriculture, carried out since 1866, is of limited value, and all other sources refer to only limited areas or points in time.

Numerous attempts have been made to estimate changes in the extent of lowland grasslands, using the 1930s or the immediate post-war years as a baseline. Fuller, Barr and Marais (1986) used, among other sources, the published and manuscript data gathered by agriculturalists, whose main interest was the agricultural value or quality of the grasslands, namely the Davies grassland survey of Wales, 1934–6, Stapledon survey of England, 1938–40, and further surveys of 1947 and 1959, Green's survey of 1970–2, and the National Farm Study, 1947–56. Bearing in mind the problems of comparing such disparate data, it appears that the area of permanent pasture, rough grazings and leys in England and Wales declined from 7.8 million ha in 1932 to 4.8 million ha in 1984, a loss of 39 per cent. Of greater significance for wildlife however, may be the decline in unimproved or rough grasslands, from 7.2 million ha in 1932 to 0.6 million ha in 1984, a decline of 92 per cent.

For some parts of Britain, the use of baselines in the 1930s, or indeed even the late 1940s, may give a misleading impression. The early 1930s were a time of acute agricultural depression, and the area of grassland may have been exceptionally large. If the aim is to gain an insight into the arable/pasture balance in times of relative prosperity, such as experienced in the decades of the 1950s, 1960s and 1970s, a more representative baseline might be the mid-nineteenth century, the period of so-called 'High Farming' (Collins, 1985).

The first Land Utilisation Survey of the early 1930s, directed by Dudley Stamp, found that the greater part of Salisbury Plain and other chalk and limestone soils was under some kind of grass cover — some naturalists and early ecologists regarded the land as having been pasture since time immemorial. This was, however, to overlook the extent of ploughing in the mid-nineteenth century, when, for example, a French topographer described how 'these immense tracts were formerly used only for sheep pasture, but the high price of corn caused them to be converted to arable' (Lavergne, 1855). Of the present-day 2,750 ha of grassland on the Porton Ranges, on the Hampshire–Wiltshire border, over three-quarters were under the plough at the time of the first large-scale mapping by the Ordnance Survey between 1856 and 1885 (Wells et al., 1976). As one agricultural commentator remarked, it was to be feared that many farmers had been tempted by the high prices for grain to borrow a large amount of capital in the full expectation that these prices would be permanent (Caird, 1852). In the event, there was a substantial fall in wheat prices from the mid-1870s onwards, largely as a result of the import of large quantities of cheap American grain. Although much of the downland arable reverted to grass, the landscapes of former times could not be recreated. Many years were needed 'to restore the old turf which the nibbling of countless generations of sheep had produced' (Sheail, 1986).

In the longer term, perhaps the greatest decline has been experienced among those grasslands defined by Tansley (1939) as neutral grasslands. These include a large range of grassland types developed on soils which are neither markedly alkaline nor acidic. They are found mostly on clays and loams, and include wet alluvial meadows, washlands, flood meadows, water meadows, pastures overlying ridge and furrow. It is estimated that about 98 per cent of the fen pastures that made up the greater part of the East Anglian Fens (an area of 3,380 sq km) in the mid-seventeenth century had been drained by 1939, and that a further 4,000 ha were reclaimed during and just after the war (Cadbury, 1984). Only 1 per cent now remains as a fragmented and much drier relic. Of the 6,094 ha set aside as washlands (safety valves in the drainage system), half is now arable (Thomas et al., 1981).

Large-scale losses of neutral grasslands are clearly not something new — what is unprecedented is the rapidity with which they have been converted to arable in some parts of the country since the last war. There have been only five occasions when field-by-field land use data were recorded for the entire area of Romney Marsh, an area of 15,800 ha in Kent and Sussex (Figure 20.1). The first occasion was as part of the Tithe Commutation Survey of 1837–44, when 88 per cent of the farmland was described as pasture and meadow. Eighty per cent of the land was described as pasture in the first large-scale survey carried out by the Ordnance Survey in the 1870s. A similar proportion was recorded as grassland in the annual census made by the Board of Agriculture from the late 1860s onwards for those parishes wholly within the Romney Marsh. Over 90 per cent of the Marsh was described as grassland in the first Land Utilisation Survey of the early 1930s. The proportion under arable in the parishes wholly within the Marsh rose from 9 per cent to 37 per cent during the 1939–45 war; 42 per cent of the area was recorded as arable or temporary grass in the Second Land Use Survey of about 1960. A further marked change occurred in the late 1960s, when, according to the census, the area under permanent grass declined from 43 per cent to 31 per cent between 1965 and 1969. Only about a third of farmland was classified as permanent grassland in the early 1980s, a similar proportion to that recorded by satellite imagery and ground-truthing in 1985/6 (Sheail and Mountford, 1984).

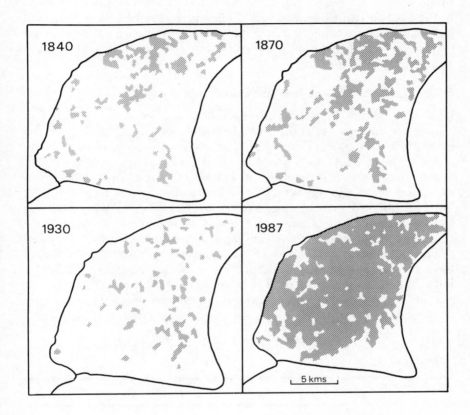

Figure 20.1 Changes in the extent of arable land (hatched) in the Romney Marsh

he exact timing of the losses in grassland area varies. The pastures of the coastal marshlands of east Essex declined from 11,749 ha to 2,083 ha, a reduction of 82 per cent, over the period 1938 to 1981 (Williams and Hall, 1987). Much of the loss (92 per cent) occurred as a result of the wartime conversion of grass to arable, and during the 1970s, when cereal prices were particularly high. In the case of the Idle–Misson Levels on the Nottinghamshire–Humberside border, much of the remaining grassland disappeared in the late 1970s, following the installation of a pump-drainage scheme. About 80 per cent of the grasslands, recorded in the Second Land Use Survey of the early 1960s, had been ploughed by 1983. The area converted to arable was seven times greater than that recorded as arable in the early 1960s and as grassland in 1983 (Mountford and Sheail, 1985).

Intensification of use of lowland grasslands

It has to be stressed that the wildlife populations of even long-established grasslands may undergo considerable change as a result of shifts in the intensity and character of management. Taking the country as a whole, there has been a large increase in the application of inorganic fertilisers (Table 20.1) and in the use of herbicides to control (eliminate) dicotyledonous weeds, which usually leads to the destruction of most dicotyledonous species. The composition of many grasslands has been affected by changes in the soil-water regime, brought about by under-drainage and mole-drainage, and improvements to arterial drainage. Changes in stocking rates, or the type of animal grazed, may have had deleterious effects on the traditional and distinctive grassland flora. The removal of animals altogether (as happens increasingly where grasslands become very fragmented in an otherwise arable landscape) soon leads to invasion by shrubs and the coarser grass species.

Table 20.1 The growth of nitrogen usage on leys and permanent grass in England and Wales, 1962–82

	Leys		Permanent grass	
g	% area treated	Kg per ha	% area treated	Kg per ha
1962	85	42	37	20
1966	69	67	44	31
1970	84	107	57	51
1974	86	133	63	69
1978	—	149	—	87
1982	93	186	77	96

Source: O'Connor and Shrubb, 1986.

Whilst there are virtually no quantitative data on their individual significance, it is the consensus view among experienced observers that these various factors, particularly when acting in combination with one another, have brought about a reduction in floristic and faunistic diversity, leading, in some instances, to the extinctions of species. Certainly, in the course of laying down a baseline from which to record future changes in plant life in western Dyfed, the Institute of Terrestrial Ecology came across numerous instances of the localised impact of changes in grass and animal husbandry, and drainage practices, in the valleys of the Upper Gwaun and Western Cleddau, and in the Castlemartin Corse and Ritec valley (Mountford and Sheail, 1986). In another survey, similarly carried out for the Nature Conservancy Council, the Institute found that half of the 240 sites, selected for survey on the basis of stratified random samples of different soil types in the Somerset Levels and Moors, had experienced some kind of agricultural improvement, for the most part by the application of fertiliser or herbicides. The pastures on such sites were composed mainly of species widespread and abundant in lowland grasslands generally. The regional and nationally rare species were largely confined to the banks of the watercourses. These refugia had been less directly affected by applications of fertiliser and herbicide (Mountford and Sheail, 1984).

Effects of changes in farming on the biota of lowland grasslands

Do these changes in the extent and character of grasslands matter? Despite their occupying so small a proportion of lowland Britain, semi-natural grasslands support about 550 species of flowering plants (about a quarter of the total British flora). Eighty-one of the species may be considered rare or endangered, either on account of their restricted geographical distribution in Britain or on the basis of their occurring in less than fifteen of the 2,900 10 km² squares comprising the National Grid (Table 20.2).

Chalk and limestone grasslands are particularly rich in rare species and those of high phytogeographic interest (fifty of the eighty-one species identified in Table 20.2 are largely confined to this habitat). The Moon carrot (*Seseli libanotes*), Monkey orchid (*Orchis simia*) and Military orchid (*Orchis militaris*) are now found at three or less sites in Britain.

Whilst changes in farming practice are by no means the only threat to species survival, the evidence available indicates that they are by far the most significant. Ratcliffe (1984) listed twenty-three characteristic vascular plants which he considered to have declined in parallel with the reduction of chalk grassland. The decline of some of these plants is chronicled in Table 20.3 and Figure 20.2, which are based on records-published in the Atlas of the British flora (Perring and Walters, 1962) and special surveys made for the Nature Conservancy Council. The decline may be worse than the maps suggest in the sense that there may

Table 20.2 Some rare and endangered plants of lowland calcareous grassland, calcareous scrub, neutral grasslands and acidic heaths and grassland

Calcareous grassland		Calcareous scrub	Neutral grassland	Acidic heath and grassland
Acerus anthropophorum	Koeleria vallesiana	Buxus sempervirens	Carex filiformis	Cicendia filiformis
Allium sphaerocephalon	Linum anglicum	Carex digitata	Carum verticillatum	Crassula tillaea
Ajuga chamaepitys	Liparis loeselii	Cotoneaster integerrimus	Fritillaria meleagris	Deschampsia setacea
Antennaria dioica	Ononis reclinata	Daphne mezereum	Gagea lutea	Erica ciliaris
Arabis stricta	Ophrys fuciflora	Lithospermum	Leucojum aestivum	E. vagans
Aster linosyris	O. sphegodes	purpurocaerulum	Oenanthe pimpinelloides	Gentiana pneumonanthe
Bunium bulbocastanum	Orchis militaris	Melampyrum cristatum	O. siliafolia	Lycopodium inundatum
	O. simia	Ornithogalum pyrenaicum		Lobelia urens
Carex ericetorum	O. ustulata		Platanthera bifolia	
C. humilis	Polygala amara			Ophioglossum lusitanicum
C. ornithopoda	P. calcarea		Ranunculus	
Cirsium tuberosum			ophioglossifolius	Pilularia globulifera
	Phyteuma tenerum		Scorzonera humilis	Rhynchospora fusca
Dianthus deltoides	Potentilla tabernaemontani		Selinum carvifolia	
D. gratianopolitanus	Pulsatilla vulgaris			
Epipactis atrorubens			Trifolium ochroleuchon	
	Salvia pratensis			
Gdium pumilum	Senecio integrifolius		Vicia orobus	
Gentianella germanica	Seseli libanotes			
G. anglica	Silene nutans			
	Teucrium botrys			
Helianthemum apenninum	Thalictrum minus			
H. canum	ssp. minus			
Herminium monorchis	Thesium humifusum			
Himantoglossum	Thlaspi perfoliatum			
hircinicum	Trinia glauca			
Hornungia petraea				
Hypochoeris maculata	Veronica spicata			
	ssp. hybrida			
Iberis amara	V. spicata			
	ssp. spicata			

Table 20.3 The decline in the distribution and abundance of five grassland species in England over the past 150 years

	Presence in 10 km² squares		
	Pre-1930	1930–76	1986
Orchis ustulata	146	66	34
Ophrys sphegodes	46	18	10
Ajuga chamaepitys	22	16	9
Fritillaria meleagris	116	15	15
Pulsatilla vulgaris	62	27	20

now be only a few individuals in some of the 10 km² squares, whereas once they may have contained separate populations of hundreds of individuals. A particularly detailed study has been made of the Pasque flower (*Pulsatilla vulgaris*). Taking thirty-two of the sites in England where the species has been recorded, it was found that ploughing had caused the extinction of the plants on twenty-five sites, and quarrying and building had destroyed the habitat of a further seven sites (Wells, 1968). Since the study was made, *Pulsatilla* has been lost from a further six sites as a result of the abandonment of traditional (grazing) management on three sites, the introduction of cattle grazing and fertiliser applications (two sites), and motor-cycling activities (one site).

Turning to neutral grasslands, the disappearance of the old-style hay meadows and of pastures free from artificial fertiliser has been reflected in a significant decline of once common species, such as the cowslip (*Primula veris*) and Meadow buttercup (*Ranunculus acris*). Among the plants which may have always been uncommon, Ratcliffe (1984) listed thirteen as now rare or endangered. It is far from easy to quantify such trends — not least because botanists in the past rarely gave precise details of the locality and abundance of all the species present — common and rare. Nevertheless, evidence of the decline of *Orchis morio* may serve as an indicator of what has happened to other, less conspicuous, species with which it is associated. Once a common plant of chalk and the calcareous clays of the Midlands and East Anglia, Dony (1976) recorded the orchid in nineteen tetrads within Bedfordshire prior to 1970; it is now found in only two. In his *Flora of Suffolk*, Simpson (1982) wrote of how the orchid was formerly frequent in old pastures and semi-scrub, but was 'now rare and exterminated in many localities'. In Kent, Philps (1982) found it had become very scarce on the chalk downland and damp meadows, where it had, until recently, been plentiful.

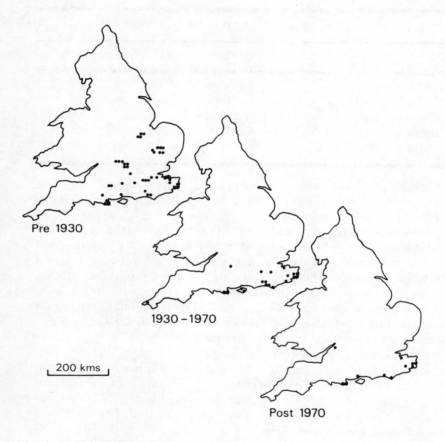

Pre 1930

1930–1970

200 kms

Post 1970

Figure 20.2 Changing distribution of Early spider orchid (*Ophrys sphgegodes*) in England

Drawing on the data derived from published flora, archival sources and fieldwork in the 1980s, attempts have been made to assess the numbers of species showing a change in abundance. The Institute of Terrestrial Ecology has carried out such an exercise for three areas (Table 20.4). Whilst there are significant differences, both within and between

Table 20.4 Numbers of plant species showing a change in abundance over the period 1880–1985

	Romney Marsh				Somerset levels and moors				Idle/mission levels			
	Increase in abundance and/or distribution		Decrease in abundance and/or distribution		Increase in abundance and/or distribution		Decrease in abundance antl/or distribution		Increase in abundance and/or distribution		Decrease in abundance and/or distribution	
	Marked	Slight	Marked	Slight	Marked	Slight	Marked	Slight	Marked	Slight	Marked	Slight
	5	20	55	42	11	21	43	58	3	5	75	47
Totals	25		97		32		101		8		122	

Table 20.5 Changes in range of butterflies over the past 150 years

Extinctions	Major contraction in range	Contraction and partial re-expansion	Contraction and equal or greater re-expansion	Major expansion of range
Larger copper (1851)	Chequered skipper	Comma	Orangetip	—
Mazarine blue (1887)	Silver-spotted skipper*	Speckled wood	White admiral	
Black-veined white (1925)	Wood white	Wall	Peacock	
Large blue* (1979)	Brown hairstreak			
	Silver studded blue			
	Small blue*			
	Adonis blue*			
	Duke of Burgundy			
	Purple emperor			
	Large tortoiseshell			
	Small pearl-bordered fritillary			
	Pearl-bordered fritillary			
	High-brown fritillary			
	Dark-green fritillary			
	Silver-washed fritillary			
	Marsh fritillary*			
	Heath fritillary			
	Marbled white*			

*Grassland species.
Source: Heath, Pollard and Thomas, 1984.

the areas surveyed, there is everywhere a preponderance of losses over increases. In the Idle–Mission Levels, where the greatest proportion of land is now cultivated, only eight species have shown signs of increasing their range and level of abundance, compared with 122 species that have become less common than they were in 1880. Even in the Somerset Levels and Moors, where habitat changes have been less extensive, 101 species have undergone a decline — a marked decline in forty-three cases (Mountford and Sheail, 1987).

The loss of plant species will lead automatically to the loss of those animals dependent on them as a food source, for example the phytophagous insects. Whilst the exact timing might vary, the records of naturalists point to a decline in the range and abundance of some butterfly species, which has accelerated in recent times (Table 20.5). The Common blue (*Polyommatus icarus*), for example, has become much less common as unimproved pastures, containing Bird's-foot trefoil (*Lotus corniculatus*) and Black medick (*Medicago lupulina*), its principal food plants, have become scarce. The reasons for the decline of other butterfly populations, such as the Small blue (*Cupido minimus*), Adonis blue (*Lysandra bellargus*), Marsh fritillary (*Euphydras aurinia*) and Marbled white (*Melenargia galathea*) are less clear. Whilst the destruction of the habitat will destroy the food plants, there may be other, contributary factors. These may include the cessation of traditional management practices, and the increased fragmentation of suitable grasslands, making it difficult for an interchange of populations to take place between otherwise isolated islands of habitat.

Birds use grasslands as feeding sites and, in some instances, for nesting. In a population study in Jutland, it was found that the numbers of Yellow wagtail (*Motacilla flava*) decreased in a seven-year period by over 80 per cent where grassland was converted to arable. The species that had the greater part of their territories within the grasslands suffered the most — the population of whinchats (*Saxicola rubetra*) fell by only 11 per cent (Moller, 1980). In the English counties of Sussex and Wiltshire, there was a dramatic decline of wheatears (*Oenanthe oenanthe*) as a result of wartime and post-war ploughing of the chalk downs — the remaining pairs were confined to the steeper slopes, where grasslands survived. There was a further decline following the outbreak of myxomatosis in 1954; there were too few rabbits to maintain the closely-cropped sward favoured by the wheatears (Shrubb, 1979; Buxton, 1981).

The effects of changes in grassland management on bird populations are often so subtle as to make it difficult to monitor them. On a Hertfordshire farm, skylarks (*Alauda arvensis*) appeared to prefer young grass leys, usually shifting their territories once the leys were more than three years old. On a farm in Sussex, where only one-year leys were grown, the skylarks seemed to 'follow' them round the farm (O'Connor and Shrubb, 1986). Cadbury (1980) attributed the decline of the corncrake (*Crex crex*) to changes in the timing of grass cutting, associated with the change over from hay-making to silage-making. Corncrakes lay their eggs in June, and the chicks hatch 15–18 days later. Previously, there was time for the chicks to be led to the undisturbed marshes. Silage-making takes place so early as to destroy the eggs in the nest.

Considerable attention has been focused on the direct and indirect effects of flood alleviation and land drainage on birds of wet grasslands (Cadbury, 1984). With better drainage, it is possible to graze higher densities of farmstock. There is a greater risk of the nests of such waders as redshank (*Tringa totanus*) and lapwing (*Vanellus vanellus*) being trampled. Although the number of nests lost as a result of one day of grazing by twenty animals may be no greater than that arising from twenty days of grazing by one animal, Dutch research has shown that the impact on the bird-breeding populations is far greater (Beintema, 1982). Whereas previously, those birds which had lost clutches soon made new nests and successfully reared the chicks, such adaptation is no longer possible where drainage improvements make it possible to keep cattle at higher densities throughout the season. Snipe (*Gallinago gallinago*) are a particularly threatened species. The bird requires tussocky grassland for nesting. As stocking rates rise, not only will there be a higher risk of trampling, but the removal of the tussocks will expose the nests to greater predation.

Research requirements

The future status and character of grasslands are central to the wider debate as to the importance which should be accorded to wildlife on farmland. As the proceedings of such gatherings as the large Symposium convened by the Institute of Terrestrial Ecology at Monks Wood in 1984 demonstrated, it is not enough to make bland statements on the need to reconcile modern farming with wildlife (Jenkins, 1984). Detailed objectives have to be set out as to what are the priorities for managing grasslands as an agricultural resource, or for the conservation of individual forms of wildlife. Once the prime use of the land has been decided, a range of secondary uses may be identified, which are compatible with the prime use and with one another.

If such complex decision-making is to be carried out competently, far more remains to be discovered as to the likely impact of combinations of land use and management on individual species and communities. A particularly exciting dimension is to discover how far techniques might be devised for creating species-rich habitats *de novo*, on land perhaps that is no longer required for cereals or other forms of intensive crop production.

Never before has there been such enthusiasm to take nature conservation so explicitly into account as part of farming operations. The question is whether such enthusiasm will be matched by the provision of resources to carry out the research necessary for devising the relevant management strategies and the practical guidelines which the man on the ground will need if there is to be any chance of turning the widely-canvassed aspirations into solid achievement.

References

Beintema, A.J. 1982. Meadow birds in the Netherlands. *Rijksinstituut voor Natuurbeheer — Rapport* (Research Institute for Nature Management Report), 1981. 86–93.

Buxton, J. 1981. *The Birds of Wiltshire*. Wiltshire Library and Museum Service, Trowbridge.

Cadbury, C.J. 1980. The status and habitats of the Corncrake in Britain 1978–9. *Bird Studies, 27*, 203–18.

Cadbury, C.J. 1984. The effects of flood alleviation and land drainage on birds of wet grasslands. In *Agriculture and the Environment*, Ed. Jenkins, D. Institute of Terrestrial Ecology, Cambridge. 108–16.

Caird, J. 1852. *English Agriculture in 1850–1*. London.

Collins, E.J.T. 1985. Agriculture and conservation in England: an historical overview, 1880–1939. *Journal of the Royal Agricultural Society, 146*, 38–46.

Dony, J.G. 1976. *Bedfordshire Plant Atlas*. Museum and Art Gallery, Luton.

Fuller, R.M., Barr, C.J. and Marais, M. 1986. Historical changes in lowland grassland. (Natural Environment Research Council contract report to the Nature Conservancy Council. Abbots Ripton: Institute of Terrestrial Ecology. Unpublished).

Heath, J., Pollard, E. and Thomas, J.A. 1984. *Atlas of Butterflies in Britain and Ireland*. Penguin, Harmondsworth.

Jenkins, D., Ed., 1984 Agriculture and the environment. Institute of Terrestrial Ecology symposium 13. (Institute of Terrestrial Ecology, Cambridge).

Lavergne, L. de. 1855. *The Rural Economy of England, Scotland and Ireland*. Edinburgh.

Moller, A.P. 1980. The impact of changes in agricultural use on the fauna of breeding birds: an example from Vendsyssel, North Jutland. *Dansk Ornithologisk Forenings Tidsskrift, 74*, 27–34 (Danish, with English summary).

Mountford, J.O. and Sheail, J. 1984. Plant life and the watercourses of the Somerset Levels and Moors. (Natural Environment Research Council contract report to the Nature Conservancy Council. Abbots Ripton: Institute of Terrestrial Ecology. Unpublished).

Mountford, J.O. and Sheail, J. 1985. Vegetation and changes in farming practice on the Idle/Misson Levels. (Natural Environment Research Council contract report to the Nature Conservancy Council. Abbots Ripton: Institute of Terrestrial Ecology. Unpublished).

Mountford, J.O. and Sheail, J. 1986. The Pembrokeshire valleys: a baseline for recording future changes in plant life. (Natural Environment Research Council contract report to the Nature Conservancy Council. Abbots Ripton: Institute of Terrestrial Ecology. Unpublished).

Mountford, J.O. and Sheail, J. 1987. Land Drainage, Habitat Change and the Grazing Marshes of Lowland Britain. Nature Conservancy Council, Peterborough, in press.

O'Connor, R.J. and Shrubb, M. 1986. Farming and Birds. Cambridge University Press, Cambridge.

Perring, F.H. and Walters, S.M. 1962. Atlas of the British Flora. Nelson, London.

Philps, E.G. 1982. Atlas of the Kent Flora. Kent Field Club, Maidstone.

Ratcliffe, D.A. 1984. Post-medieval and recent changes in British vegetation: the culmination of human influence. New Phytologist, 98, 73–100.

Robinson, M.A. and Lambrick, G.H. 1984. Holocene alluviation and hydrology in the upper Thames basin. Nature, 308, 809–14.

Sheail, J. 1986. Grassland management and the early development in British ecology. British Journal of the History of Science, 19, 283–99.

Sheail, J. and Mountford, J.O. 1984. Changes in the perception and impact of agricultural land improvement: the post-war trends in the Romney Marsh. Journal of the Royal Agricultural Society, 145, 43–56.

Shrubb, M. 1979. The Birds of Sussex. Phillimore, Chichester.

Simpson, F.W. 1982. Simpson's Flora of Suffolk. Suffolk Naturalists' Society, Ipswich.

Tansley, A.G. 1939. The British Islands and Their Vegetation. Cambridge University Press, Cambridge.

Thomas, G.J., Allen, D.A. and Grose, M.P.B. 1981. The demography and flora of the Ouse Washes, England. Biological Conservation, 21, 197–229.

Wells, T.C.E. 1968. Land-use changes affecting Pulsatilla vulgaris in England. Biological Conservation, 1, 37–43.

Wells, T.C.E., Sheail, J., Ball, D.F. and Ward, L.K. 1976. Ecological studies on the Porton Ranges: relationships between vegetation, soils and land-use history. *Journal of Ecology*, *64*, 589–626.

Williams, G. and Hall, M. 1987. The loss of coastal grazing marshes in south and east England, with special reference to East Essex, England. *Biological Conservation*, *39*, 243–53.

Management of Wetlands

Introduction

I.M. Tring, Ministry of Agriculture,
Fisheries and Food. London

It is clear that over the last few decades, important wetland has been lost or irreversibly damaged for a variety of reasons. Pressure from government for increased production supported by grants for drainage has been an important influence but greater public awareness has begun to redress the balance; but we must not be complacent. Some participants expressed the wish to have the back-up of further legislation and financial support to provide protection for wetlands.

Summary report

At the Workshop Dr Culleton aptly stated that all concerned face a 'multifaceted challenge'. Drainage (both field and arterial) and associated ancilliary work are considered either a primary or the single greatest threat to wetlands. However, agricultural management systems, urban and industrial development, flood protection schemes, pollution from a variety of sources, recreation and tourism are all significant pressures.

From an engineering viewpoint it is certainly possible to lessen the destructive impact of both improvement and maintenance works by sensitive scheme design, careful execution and continuing education. The Workshop clearly demonstrated the need to bring agriculture and environmental interests into closer harmony.

In the past, drainage appeared to have been the most widely practised form of agricultural improvement particularly in Northern Europe. There was a levelling off in the early 1980s for most Northern European countries (except France) and in the United Kingdom, field drainage works have declined by more than 50 per cent.

At present it is far from clear whether drainage represents the best investment for raising farm incomes and clearly more careful appraisal of the economic value and the merits of alternative land use is overdue. The current state of UK research on grassland drainage indicates the practice to be economic in only two years out of five.

Research and development

Participants were keen to fill the gaps in present knowledge but it was clear that they lacked information on the current state of knowledge in the wider Community. The Workshop highlighted the need to: Increase knowledge on environmental impact assessments; Develop present ideas; Give urgent attention to the problem of eutrophication; Pay greater attention to landscape.

International Commission for Irrigation and Drainage (*ICID*)

ICID has set up a Working Party on the environmental impacts of irrigation, drainage and flood control works. Its long-term objective is to draw together information currently available on environmental effects and to provide guidance to project designers and managers in identifying and minimising environmental degradation. Those wishing to provide an input should contact their respective national representative.

Recommendations

1. Set up a mechanism whereby scientists and others are aware of relevant R & D to provide a vehicle for co-operation between Member States. This may be achieved in part through similar workshops.
2. Investigate the economic implications of alternative land use, targeted on environmental objectives. This applies to wetlands and to areas adjoining wetlands.

Applied ecological research on the conservation of wet grasslands in relation to agricultural land use in Flanders (Belgium)

E. Kuijken, Instituut voor Natuurbehoud, Hasselt.

Traditional farming in western Europe created and maintained highly interesting semi-natural landscapes for many centuries, notably many types of heathlands and grasslands. And in the wetlands moderate usage did not place an excess burden on their ecosystem. As a result, except in the woodlands, a relatively stable balance between man and nature existed. During the last few decades, however, the development of modern agriculture has caused growing environmental problems on both spatial and qualitative levels (Grootjans, 1985; Grootjans and Klooster, 1980).

In this chapter, we discuss relations and conflicts between agricultural land use and nature conservation, and the role of ecological research in an attempt to reach a balance between productivity and maintenance of traditional semi-natural landscapes, notably wet grassland ecosystems in polders and river valleys.

Wet grasslands: ecological characteristics and importance

Origin

Traditional use of some stages in the succession of wetlands gave way to the development of grassland ecosystems: summer mowing of reed and sedge marshes, hay-making in river valleys, heathland clearing, and cattle or sheep grazing in saltings and other tidal marshes. For many generations this extensive land use remained largely unchanged, which enabled an adapted ecosystem development with a high degree of diversity and internal stability. These semi-natural landscapes consist of spontaneous flora and fauna elements, whereas the structure of the vegetation is mainly determined by the (moderate) human land use. These grasslands are also characterized by quite dynamic environmental factors, notably seasonal fluctuations of water levels and related changes in nutrient input. Interesting variations occur in geomorphology (micro-relief) and soil composition (organic or mineral sediments), as these wet grasslands form a part of the macro-gradient of the landscape. They are indeed situated between the original, real wetlands (if still existing) and more intensive agricultural land.

Typology

Classification is mainly based upon four groups of features:

(1) *abiotic factors*: geomorphology, soil, hydrology, water quality; differrences and gradients exist between salt or fresh, permanent or temporary wet, eutrophic or oligotrophic;
(2) *land use and presence of artefacts*: traditional hay making, grazing, use of fertilisers, amount of drainage, ploughing up and reseeding; presence of artefacts such as dikes and ditches can contribute in the total habitat diversity and vegetation present;
(3) *vegetation*: reflecting the above mentioned features, the traditional phytosociological subdivisions illustrate the great variety and ecological diversity that exist within the term 'grasslands' in Flanders most types belong to the alliances Calthion, Filipendulion, Molinion, Lolio-Cynoturion and Arrhenatherion;
(4) *fauna*: avifauna is most striking; a distinction is to be made in the function of grasslands for breeding birds (some critical species), or their role as feeding grounds for large concentrations of wintering/migrating waterbirds: the occurrence of invertebrate fauna largely depends on the structure of the vegetation and of the habitat as a whole (shelter, hibernation).

Ecological importance

The importance of wet grasslands within the scope of nature conservation can be summarized by mentioning the following functions: genetic reservoir, ecological regulation of hydrological systems, auto-purifying capacity of flooding water, economic use (even if not improved), historic testimony of landscape development by man, aesthetic value (Coupland, 1979).

Moreover, some grassland complexes still cover relatively large open areas, which hold more actual and future capacities for optimal ecological functioning than smaller, isolated habitats. In these grassland areas examples of the landscape, an ecological network of ditches, hedgerows, dikes, verges, slopes, depressions, pools and rows of trees is still to be found (connectivity functions).

In order to conserve these intrinsic ecological values, it is important to continue traditional land use where possible, even with modern technology. In particular the maintenance of high water levels and natural fluctuations, poor or at most moderate manuring, late mowing dates (not before mid-June), extensive grazing, no ploughing and reseeding and especially a fixed use in a pattern of pastures, hayfields and some mixed meadows (Van Duuren et al., 1981).

Threats and conservation measures

Agricultural improvements versus nature conservation

Several developments cause a general decrease of the ecological characteristics and functioning of rural landscapes: urban and industrial expansion, agricultural and para-agricultural developments, (sub)urbanisation of rural villages, extension of outdoor recreation, construction of roads. Especially in Flanders with a population density of almost 420 inhabitants per square kilometre, the land is split up in a relatively haphazard chaotic way. Apparently, the laws on physical planning came too late (1962, 1972) to remove the negative effects of the past and to convince the public that future land use planning needs stronger measures. Besides the spatial threats mentioned, a series of major qualitative dangers are related: pollution, acidification, eutrophication, over-use of fertilisers, disturbance of water supply, drainage and other changes.

In the last few decades, agriculture has had to give up a large area of highly productive land for the above mentioned developments. In Belgium 265,000 ha were lost between 1960 and 1984, of which 60 per cent was in Flanders. As a result interest has increased in the remaining less cultivated semi-natural areas in an attempt to compensate for these losses. Here the effects of modern agricultural improvement cause even more conflicts than in traditional productive areas. The following steps illustrate the mechanism:

—lowering the water level in valleys and polder depressions: building pumping stations, rectification of rivers, rivulets and brooks, creation of artificial water reservoirs for stocking water 'surplus' to avoid larger inundations (mainly in winter);

—on these drained soils, earlier access in spring with heavier machines is possible, enabling earlier and increased use of fertiliser (plus use of herbicides); after only one season, a drastic decline in the original

diversity results, with several sensitive plant species disappearing;
—old permanent grasslands (especially hayfields) are then ploughed
and reseeded with highly productive mixtures of grass species and
even more fertilisers are added;
—as a result of decreased flooding, better drainage and nutrient
input, earlier and more frequent mowing (silage grass) becomes a
general practice; this causes conflict with breeding birds and gives
plant species no chance to seed; most critical species only survive
in the field margins and along ditches;
—due to European milk quotas, more and more grasslands are
irreversibly changed into fields (between 1977 and 1984, 3,570 ha
of grasslands disappeared each year); besides the traditional crop
of potatoes, maize crops are preferred at present, because of the
high amount of animal dung that can be dumped here (surplus
available from intensive bio-industries);
—as a result, a tremendous over-eutrophication has occurred during
the last ten years, notably in the northern and western part of
Flanders; this causes many negative side effects even in the best
protected nature reserves (ruderalisation); further, sport fisheries
and catchment of drinking-water became quite impossible.

Most of these developments form part of reallotments and hydraulic
improvement plans, which still aim at increased productivity. In the
context of the European Community's agricultural surpluses, it is
debatable whether the subsidies concerned are still justified. The
psychological aspect, however, is that farmers expect a further
income increase, because of the large investment they have to make.
Lowering the rent on marginal land and revising the fiscal system
could reduce the pressure to improve semi-natural grasslands. In
particular, EC subsidies for farming in areas with a natural handicap
(e.g. mountains) should be extended to the last inundable valleys in
Flanders. In 1960 flooded areas still covered 35,000 ha; this has been
reduced to less than 10 per cent of that area in 1987.

Plantations

In accordance with our argument that continuity of traditional land
use can be of importance in maintaining some ecological master
factors, the problem of grasslands being abandoned also exists to some
extent. For many decades, wet grasslands of river valleys in Flanders
have been planted with poplars. Thus, they lost their agricultural
importance, but this also causes conflicts with nature conservation.
In some cases, restoration of rich herbaceous vegetation has been
realised in nature reserves.

Conservation measures

Unlike real wetlands, many of which have a status as nature reserves, the last few large complexes of wet grasslands lack serious legal protective measures. The sectorial physical planning maps in Flanders, however, include some appropriate designations:

(1) *R- and N- zones*: concern nature reserves and sites with special conservation interest, representing respectively 1.1 per cent and 5.7 per cent of the Flemish territory. In principle they are designated for the maintenance (and restoration) of the natural environment. Some of the most important grassland areas occur in these categories, approximately 1.0 per cent of the territóry.

(2) *V- zones*: 'valley-areas' or 'agricultural areas with ecological importance'. Here agriculture can continue as long as the natural environment of flora and fauna is not disturbed. Most zones concern wet grasslands that are inundable in winter, or were until recently. Some grasslands surrounding (wetland) reserves as bufferzones also have these designations. In Flanders they amount to 1.1 per cent of the land area.

In practice, however, these official protective designations have little effect because most (farm) land is privately owned. Agriculture in particular does not readily accept restrictions, even in N- and R- zones. In V- zones there is even less respect for environmental needs. There are several examples of reallotments that include highly interesting semi-natural grasslands (R-, N-, V- zones) and even wetlands, that are the subject of drainage or lowering of water-levels resulting from the building of new pumping stations. As a result, private farmers feel stimulated to increase their productivity, the more so as they have to pay a part of the reallotment costs.

In other cases the restrictions of designations remain theoretical, because the alteration of vegetation can be done without asking legal permission. Thus ploughing old grasslands, drainage, filling up depressions, digging ditches, over-use of fertilisers became common practice by private farmers: this is called 'autonomous development', where control from government or conservation authorities does not exist. In ongoing reallotments, there is at least a landscape plan (as, however, advice only) that normally also includes landscape-ecological inventories, evaluations and propositions for conservation.

Applied ecological research

In order to introduce environmental needs into agricultural management, applied ecological research is needed. It has to include formulations of aims, methods and limits of nature conservation and identification of possible conflicts, based upon understanding of ecological mechanisms. This must lead to practical advice concerning selection of priority habitats and areas for either mono- or multifuctional land use, taking into account the incompatibility of some developments. More detailed research is needed in view of setting up management plans. Programmes for monitoring trends in environmental master factors are necessary in order to support and adjust long-term policy (Boedeltje and Bakker, 1980).

Large-scale surveys

Landscape ecological surveys provide most relevant background information for nature conservation planning. Based upon mapping of vegetation (being the best expression of soil and (ground) water conditions) this also takes into account the implications of actual land use. Additional information on fauna can be included if sufficient and complete references exist (e.g. waterfowl numbers and distribution). A national programme of ecological mapping, started in 1978, will cover the whole Belgian territory (scale 1:25.000) (De Blust *et al.*, 1985). The manual of cartographic units or ecotopes is mainly based upon phytosociological alliances or associations (90 units) and is completed with a list of typical man-made structures in the landscape (25 units). Units are mapped either separately or as complexes (especially when linear or small elements occur or when heterogenous areas are concerned). A 'biological value' has been given to each unit, using three classes of conservation importance. This assessment is based upon the criteria rarity, biological quality (including diversity), vulnerability and replacability. The lowest class includes 50–70 per cent of the territory, the highest one 5–8 per cent, depending on regional differences. Delimitations and symbols of the units are over-printed on the topographic maps and evaluation is represented by colouring the three classes. The addition of such an evaluation was the subject of serious scientific criticism. Against this must be set the fact that this improved the use of this document as a *signal map* for land-use planners showing priorities for nature conservation (Kuijken and Heirman, 1984).

Specific research

More limited inventories are prepared on demand for some land improvement programmes. In these cases, detailed records of vegetation, using Tansley or Braun–Blanquet methods, give useful information on the presence, abundance and distribution of separate plant species and communities. This enables a more objective calculation of derived parameters (rarity) or indicative values (sensitivity, vulnerability).

In these projects, decision makers also ask for predictions of possible side-effects in order to prevent controversy. Answering this is quite difficult, as detailed insight into dose–effect relations and understanding of complex relations is seldom available. Comparisons with analogous situations can help, but detailed monitoring of ecological reactions on changes in land use is again lacking.

The best method is the combined use of the different indicative values, contributed to plant species by Ellenberg (1974). Based upon a number of records, figures for humidity, pH and nitrogen level can be calculated for critical sites, as well as their average rarity. Also the use of Londo's list of phreatophytes enables the calculation of classes in relation to dependence on ground-water regime. Based upon the relations that exist between the parameters mentioned, combinations can be used as an assessment of vulnerability of vegetation. For this purpose, a relevant number of records is needed, as well as exact data on the ecological habitat conditions (average and extreme water levels, degree of manuring, pH of soil and water).

Concluding remarks and suggestions

(1) In agricultural improvement programmes, ecological research should be introduced long before the planning is initiated and fixed; this should result in official advice, not only of small modifications required in the execution of works, but also of reconsideration and if necessary change in the original objectives. In recent years, this has been an increasing matter of concern to nature conservation societies and administration.

(2) In areas where designation maps (physical planning) and ecological evaluation maps indicate the presence of importance conservation values, agricultural improvement plans have to be excluded. When side-effects are feared as a result of interference from neighbouring developments (drainage, eutrophication), the above-mentioned advice should be combined with a right of veto. In these officially recognised zones of conservation interest, any change of vegetation, micro-relief and water table must become subject to a system of obligatory permissions, in order to prevent further uncontrolled autonomous developments.

(3) Research should be stimulated concerning technological solutions of the growing surplus of waste from bio-industries in some regions. In addition further overall hyper-trophication of the environment, which also has negative effects on agriculture itself, should be

prevented. Also higher standards on the emission of harmful products in nature reserves are needed.

(4) The area of semi-natural habitats, especially grasslands, is rapidly declining in Flanders. Only a few larger complexes remain where flooding in winter can still occur. These areas have distinct botanical or ornithological values that could justify their inclusion in the Ramsar list of wetlands of international importance (where countries are urged to give more priorities to inundable grasslands: Cagliari Conference 1981). Also their status as special protection areas within the EC directive on bird protection could contribute.

(5) Most agricultural improvements cause ecological effects that are contrary to the efforts and results of management in nature reserves. Thus viable agriculture is rarely compatible with proper nature conservation. Conservation, however, should be restricted to reserves only. Therefore, a system of gradual segregation and integration according to the characteristics and qualities for both functions is to be studied (land use planning).

(6) In the context of the socio-structural measures of the EC, it is not clear if withdrawing land from agriculture will be accepted in Flanders as a relevant solution to surpluses. In the case of wet grasslands, for instance, an integration model seems feasible when the income of individual farmers concerned does not fully depend on these 'marginal' grounds. It is to be recommended that financial support should be given to those farmers both for not improving semi-natural habitat and for integrating the maintenance of other ecologically important landscape elements (e.g. hedgerows). The growing conviction that agriculture, as well as conservation, has both historic and environmental responsibility can also be stimulated by contributing European funding to compensate income loss when traditional use of monumental old ecosystems is continued. It is hoped that in these large landscapes restoration of natural values can also be realised in a new balance with other social needs.

References

Boedeltje, G. and Bakker, J.P. 1980. Vegetation, soil hydrology and management in a Drenthian brookland. *Acta Bot. Neerl.*, 29, 509–522.

Coupland, R.T. 1979. *Grassland Ecosystems of the World: Analysis of Grasslands and their Uses* (IBP 18) Cambridge University Press, Cambridge

De Blust, G., Froment, A., Kuijken, E. Nef, I. and Verheyen, R. 1985. *Carte d'evaluation biologique de la Belgique; Texte explicatif general* Minist. de Sante Public et de l'Environnement; Institut d; Hygiene et d' Epidemiologie; Bruxelles.

Ellenberg, H. 1974. *Zeigerwerte der Gefasspflazen Mitteleuropas.* Gottingen.

Grootjans, A.P. 1985. Changes of groundwater regime in wet meadows. Doctoral dissertation, Rijksuniversiteit Groningen, Netherlands.

Grootjans, A.P. and Klooster, W.Ph Ten. 1980. Changes of groundwater regimes in wet meadows. *Act. Bot. Neerl.*, *29* 541–554.

Kuijken, E. and Heirman, J. 1984. The Biological Evaluation Map of Belgium: an applied ecological inventory program. In *Further Examples of Environmental Maps* (ed. D. Bickmore), International Geographic Union/International Cartographic Association; Joint Working Group on Environmental Atlases and Maps, Madrid (Spain). pp. 47–50.

Van Duuren, L., Bakker, J.P. and Fresco, L.F.M. 1981. From intensively agricultural practices to hay-making without fertilisation. Effects on moist grassland communities. *Vegetation*, *47*, 241–58.

22 Plant production and agricultural use of wet lowland and dry sandy soils in Denmark

L. Hansen Statens Forsogsstation, Tinglev.

Farmland in Denmark totals 2,800,000 ha, including 220,000 ha, or 7 per cent, permanent grassland. Permanent grassland exists as small areas on the coast and along the rivers. Normally these areas comprise only a small part of a farm, and are used for grazing cattle. Arable farming is possible on almost all Danish soils, and is intensively practised. The use of fertilisers is necessary. Wet lowland has to be well drained when used for arable farming. Dry, coarse, sandy soils have to be irrigated. In Denmark, there is growing public interest in changing part of the intensive arable farmland to extensive cultivation in order to protect the environment in general.

The problems and management of marginal soil in Denmark have been evaluated in at least forty technical reports. Part of the intensive cultivated areas will probably be changed to marginal farmland. Some believe that up to 20 per cent of arable farmland will be changed to marginal land in ten to fifteen years. Summaries of two of the technical reports follow.

Coarse sandy soil

Coarse sandy soils comprise about 700,000 ha or 25 per cent of the arable land. These soil types are extensive in the western part of Denmark and they are often used for mixed farming with dairy cows. Many of these areas were reclaimed from heath soil a century ago. By liming, manuring and use of fertiliser these soils have been changed

Table 22.1 Yields in a rotation of four crops: crop units per hectare, average over twenty years

		Crop units x 100		Relative yields		
Location	Untreated Fertiliser	Manure	Untreated	Fertiliser	Manure	
Askov	1923–48	10.8	39.5	36.5	27	100	92
Lundgard	1927–46	13.9	33.0	32.9	42	100	100
Studsgard	1929–44	11.2	30.4	28.6	37	100	94
Tylstrup	1927–46	22.1	42.4	40.6	52	100	96

to productive arable farming. Fertiliser treatment by the principle of substitution is necessary for economical farming. Crop yields from twenty years' experiments are shown in Table 22.1. Crops were grown on a four year rotation: rye – swedes – oats – clover. Without treatment with fertiliser the yields are very low — on grassland as low as 500–1,000 crop units.

In Denmark 400,000 hectares or 14 per cent of the total farm area is irrigated. In regions with coarse sandy soils 30–40 per cent is irrigated. With irrigation the crop yields increase 20–50 per cent, at the same time yield values from year to year are stabilised. The effect of irrigation combined with treatment with fertiliser is shown in Table 22.2. On coarse sandy soil there is no alternative farming system to intensive arable cultivation. Without treatment with fertiliser and irrigation the yield will be very low. The only use will be permanent grass of low quality, therefore management has to be subsidised by the State.

Table 22.2 Yield and relative yield after irrigation and treatment with fertiliser; average for coarse sandy soils Relative yields

	Yield		Irrigated			Not irrigated	
	hKg/ha	N	0.75N	0.5N	N	0.75N	0.5N
Barley	48	100	94	85	73	67	58
Winter barley	55	100	89	75	69	62	51
Rye	51	100	94	84	84	78	69
Wheat	60	100	90	75	62	53	42
Rape	24	100	96	93	83	79	71
Potatoes	100	100	96	86	77	73	64
Sugar-beet	140	100	95	80	82	78	66
Grass	99	100	90	75	87	78	66
Clover	90	100	96	89	82	72	59

Table 22.3 Use and yield of grass on some lowlands

Location	Area ha	Heifer number	Heifer per ha	Crop units per ha	Remarks
Nyord	237	175	0.8	590	
Skallingen	1,000	300	0.3	240	
Tipperne	500	440	0.9	560	
Varnenge	580	345	0.6	470	
Varnenge	750	900	1.2	960	Fertiliser used
Vildmose	460	1,732	3.8	3,000	Fertiliser and drainage used
Tondermarsk	1,400	4,400	3.1	2,514	

Source: Lundsmark Jensen 1987

Wet lowland

In Denmark 500–600,000 hectares or 15–20 per cent of the area is lowland. Wet lowlands on the coast and along the rivers have sandy and peaty soils. Most of these soils have been satisfactorily drained and are used for arable farming with a high yield in cereals and grass. On peaty soils the drainage system has to be renewed after twenty or thirty years. Without drainage these soils will revert from arable farming to permanent grassland. Grasslands are utilised for feeding heifers from dairy cow livestock. Surface drainage is necessary. Productivity from some lowlands is shown in Table 22.3. On drained and fertiliser-treated lowland grass yields are high. Without fertiliser yields are low. In Denmark there is a shortage of young cattle and very few sheep. Wet grasslands are of low value in the normal farming system, therefore management of such land has to be subsidised by the State.

References

Djurhuus, J. 1987. Landbrugsmaessig udnyttelse af vandlobsnaere arealer. *Teknisk rapport No. 23*, p. 52.

Jacobsen, Sv. E. and Abildskov, A. 1987. Produktion of landbrugsafgroder pa torre, sandede jorder. *Teknisk rapport No.14*, p. 214.

23 | Topsoil compaction causes ponding in grazed pastures

J. Mulqueen, Agricultural Institute, Ballinrobe.

In the wet summer of 1986 many cases of ponds as large as 0.5 ha in grazing fields were investigated mainly in the western and southern parts of Ireland. These ponds formed on free draining soils such as brown earths where ground water-tables are deep. It was notable that the ponding occurred outside the areas of impervious clayey soils known for their susceptibility to poaching. The ponding was most common on intensively stocked dairy farms but it was also found on intensive cattle, cattle-sheep and horse-cattle (stud) farms. The ponds persisted throughout the five week dry spell of September–October 1986 when only 5 mm of rain fell. This ponding and its associated effects have both agricultural and environmental implications. This chapter is a report on soil investigations carried out into the causes of ponding at Athenry Agricultural College, Co. Galway. The ponding is discussed in relation to causative factors and agricultural and environmental issues.

Literature review

Heavy grazing greatly reduces the infiltration capacity of soil (Holtan and Kirkpatrick, 1950). Their mass-infiltration curves for a range of conditions show rates of 50, 40 and 30 mm/hr for lightly, moderately and heavily grazed pastures. Lull (1964) states that compaction of the surface soil is the major factor in reducing infiltration. This takes place through a reduction in non-capillary porosity. Musgrave and Holtan (1964) have given a comprehensive treatment of infiltration and Warren et al. (1986) have reviewed the effects of trampling damage on soil hydrologic characteristics.

Table 23.1 Calculated stresses in the soil at a depth under a free standing cow

Depth (cm)	0	5	10	15	20
Stress (bar)	1.7	1.1	0.5	0.25	0.15

Source: Sohne, 1953

The hooves of grazing cattle apply high pressures to soil causing deformation of the soil and trampling the pasture (Scholefield *et al.*, 1985). The nature of the soil deformation depends on the soil moisture content. At high moisture content, plastic flow predominates while at low soil moisture content, compression predominates (Mulqueen *et al.*, 1977). At intermediate moisture contents both compression and shear are important. In a soil profile, plastic flows may predominate in the upper wetter topsoil layers and compression in the lower, drier more confined topsoil layers. The effects of compaction of agricultural soils have been reviewed (Agricultural Advisory Council, 1970; American Society of Agricultural Engineers, 1971).

Treading by cows and other cattle

An intensively grazed pasture is severely trodden by cows and other cattle. A cow of 550 kg has a hoof area of about 320 cm² exerting a deadweight load of 1.7 bar. Such a cow exerts a vertical axial load of 3.75 bar when walking (Scholefield *et al.*, 1985). A cow walks about 3 km in 24 hours treading about one sixtieth of a hectare assuming she does not retread the same area twice.

The stress distribution with depth in soil under a hoof is calculated assuming a uniformly loaded circular area (see Table 23.1). This shows that at depths of 20 cm and greater the stresses become very small. Similarly, the stresses caused by walking cows are low, below 20 cm. These calculations indicate that, excepting very soft soils, significant soil deformations by grazing cows are likely only in the upper 20 cm or so in normal and firm soils.

Investigations

Sites with ponds were surveyed and test pits were excavated. Notes were made of the density, structure, colour and smell of the soil at the various depths. Soakage tests were carried out by pouring 250 ml of water into 150 × 150 mm sumps in the bottom of test pits

Table 23.2 Physical properties of deformed Athenry Agricultural College topsoil

	Depth (cm)		
	0-5	5–10	10-15
Bulk density			
(g/cm³)	1.03	1.27	1.38
Porosity (%)	56	48	46
H₂O (% v/v)*	58	46	40
Saturation (%)	103	96	87

*Sampled 24/10/86 when topsoil was very wet.

sunk to various depths. At Athenry Agricultural College density, porosity and moisture content of the soil were measured. Estimates of permeability were made because of the special difficulties of measurements on gravelly soils.

Results

Results are presented for grazing fields at Athenry Agricultural College. Surveys showed that ponding developed in hollows as a result of surface run-off from the higher surrounding ground and low infiltration capacity in the hollows. The surface run-off was caused by topsoil deformation, which lowered the infiltration capacity to negligible values in both the high and low ground.

Test pits showed that the deformed topsoil away from and near the ponds could be divided into two layers: a surface 0–7 cm layer and a bottom 7–17 cm layer. These rested on a loose brown loam of granular structure with an estimated permeability of 500 mm/day. The surface topsoil layer was wet, soft and weak and readily peeled away from the bottom topsoil layer. It was structureless and puddled and contained virtually all the grass roots. It was dark grey in colour and smelled unpleasant (anaerobic).

The bottom topsoil layer was very dense and firm and broke away from the underlying loose soil in a horizontal slab. It was a dark grey colour, anaerobic and foul smelling. On breaking up and exposure to air it changed to a dark greyish brown colour as it oxidised. It had no visible living roots and the few hair cracks and root channels were heavily stained with rust-coloured iron compounds. This compacted state contrasted sharply with the brown to dark greyish brown colour,

Table 23.3 Rainfall and potential evapotranspiration (PE) at Galway 1985 and 1986 (mm)

Month	April	May	June	July	Aug.	Sept.	Total
1985	61	111	46	70	238	126	650
1986	74	175	84	70	146	5	554
Mean	66	70	73	79	97	111	496
PE	50	79	81	74	62	37	384

granular structure and mellow smell of uncompacted topsoil near hedgerows and fences.

The soil layer below about 17 cm was dark brown, friable and of granular to sub-angular blocky structure. Soakage tests indicated good permeability of about 0.5 m/d.

Physical properties of the topsoil (Table 23.2) confirm that the upper topsoil layer was puddled while the bottom topsoil layer was compressed and compacted. The compacted 10-15 cm layer appeared to be unsaturated possibly as a result of entrapped gas but more work is required on this aspect.

Mechanism of topsoil deformation and its effects

The stocking rate on the dairy farm at Athenry Agricultural College is 2.5 cows/ha and this results in intensive treading of the soil. The topsoil is a loam to gravelly loam with granular structure on the hillocks and a loam to silt loam with granular structure in the depressions (Diamond et al., 1965). Typical texture analysis on the hillocks are 20 per cent coarse sand, 25 per cent fine sand, 40 per cent silt and 15 per cent clay with slightly more clay in the depressions. The topsoil contains 12 per cent organic matter on average in the top 15 cm.

The grazing seasons of 1985 and 1986 were very wet (Table 23.3) with May and August extremely wet in both years. When the wet and damp surface topsoil layer was treaded by a hoof, it was deformed, flowed sidewards and bulged upwards around it. Flow planes could be easily distinguished in the soil about hoof prints. The soil was puddled giving rise to low permeability and anaerobic conditions. This resulted in reduced grass growth and ponding in hoof prints predisposed the soil to more puddling.

Directly beneath the hoof the bottom topsoil layer was drier and more confined so that it was compacted, causing the soil particles to move very close together and slide over one another. This process gave rise to a very tight strong layer, practically impervious and anaerobic. This layer appears to be chiefly responsible for surface run-off and ponding.

Topsoil compaction in relation to agricultural and environmental issues

In agriculture the main effect of topsoil compaction is ponding, poaching and shortage of feed caused by reduced grass growth, grass cover and efficiency of nitrogenous fertiliser. Topsoil compaction and ponding predispose the surface topsoil layer to subsequent poaching and grass is trampled into the ground. As a result the stocking rate had to be reduced to 1.9 cows/ha in the months of August and September 1986 at Athenry. Management difficulties increase from badly soiled cows requiring more washing and teat care. Increased soiled water must be irrigated on the soil further increasing the hydraulic loading. Bringing cows indoors in wet weather increases the feed requirements and volume of slurry. In badly poached fields, some of the bare ground became colonised by annual meadow grass (*Poa annua*).

Environmental changes include ponds in the landscape and when these are drained away by loosening the compacted soil, bare hollows which must be reseeded. Severe poaching increases the amount of bare ground and as highly productive grasses are trampled into the soil, there is increased invasion of annual meadow grass. Around the edges of the ponds there is invasion of aquatic weeds. If heavy rainfall follows on the application of slurries and fertilisers, some of these are washed into the ponds where they may give rise to polluted waters and smells. Wetland birds such as snipe (*Gallinago gallinago*) are encouraged to the very wet ground at the ponds.

The soil investigations show that certain free draining brown earth topsoils can be transformed to impervious gleyed topsoils by intensive treading by cows and cattle in wet grazing seasons. This has relevance to the stability of agricultural and environmental ecosystems. While much of the puddling and compaction can be undone by soil loosening when the soil is dry, the loosening itself does not restore the original structure. Apart from this and from the loss of production and ground restoration problems in loosening, it is not yet known if there are long-term effects especially in predisposing the soil to repuddling and recompaction.

References

Agricultural Advisory Council. 1970. *Modern Farming and the Soil.* HMSO, London.

American Society of Agricultural Engineers. 1971. *Compaction of Agricultural Soils.* American Society of Agricultural Engineers, St. Joseph, Michigan, USA.

Diamond, S., Ryan, M. and Gardiner, M.J. 1965. Department of Agriculture School Farm, Athenry, Co. Galway. *Soil Survey Bulletin, 12,* An Foras Taluntais, Dublin.

Holtan, H.N. and Kirkpatrick, M.H. 1950. Rainfall, infiltration and hydraulics of flow in run-off computation. *Trans. Am. Geophys. Union, 31,* 771–9.

Lull, H.W. 1964. Ecological and Silvicultural Aspects. In *Handbook of Applied Hydrology* (ed. Ven te Chow). McGraw-Hill, New York, pp. 6,14.

Mulqueen, J., Stafford, J.V. and Tanner, D.W. 1977. Evaluation of penetrometers for measuring soil strength. *Journal of Terramechanics, 10,* 137–51.

Musgrave, G.W. and Holtan, H.N. 1964. Infiltration. In *Handbook of Applied Hydrology* (ed. Ven te Chow). McGraw-Hill, New York, pp. 12.1–12.30.

Scholefield, D., Patto, P.M. and Hall, D.M. 1985. Laboratory research on the compressibility of four topsoils from grassland. *Soil and Tillage Research, 6,* 1–16.

Sohne, W. 1953. Quoted in Koolen, A.J. and Kuipers H. 1983. *Agricultural Soil Mechanics.* Springer-Verlag, Berlin, p.36.

Warren, S.D., Thurow, T.L., Blackburn, W.H. and Garza, N.E. 1986. The influence of livestock trampling under intensive rotation grazing on soil hydrologic characteristics. *Journal of Range Management, 39,* 491–5.

24 Reclaimed Danish lakes and inlets: agricultural value and nature interests

J. Waagepetersen, Danish Land Development Service, Viborg.

As part of comprehensive work with marginal land in 1986 carried out by the Danish Ministry of Environment, the Danish Land Development Service, Dansk Ornitologisk forening, Bio Consult and Geo Consult have mapped the position of drained lakes and inlets. Furthermore their agricultural value as well as their present and potential nature values were determined. The following is a summary of the resulting report.

Countrywide mapping

Figure 24.1 shows the position of all registered reclaimed lakes and inlets. It is estimated that an area comprising 55,000 ha of lakes and inlets has been reclaimed. In addition drainage has resulted in reclamation of the adjacent low-lying areas.

The reclamation activity was especially marked in the latter part of last century, when drainage was carried out by means of wind-driven Archimedian screw pumps. This century saw these replaced by electric pumps and an overall improvement of the drainage systems. The Danish Land Development Service has been responsible for a great many of these modernisations.

Choice of pilot-plants

The thirty-one areas, shown in Figure 24.2, were examined in detail. The examined areas comprised five areas which were relatively closely examined (Bundso–Mjelsso, Soborg So, Sem So, Henninge Nor and Bolling So). In addition twenty-six areas covering large drainage projects in the communities of Funen and Southern Jutland, as well as a large part of the projects undertaken in Viborg, were extensively examined. The average

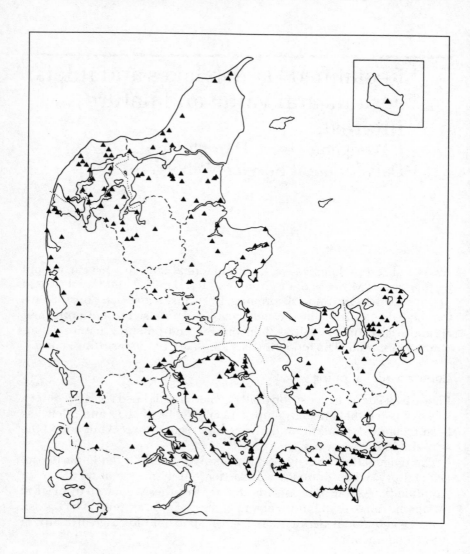

Figure 24.1. Mapping of drained lakes and bays in Denmark

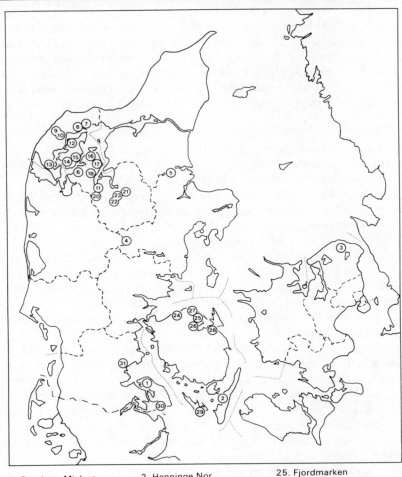

1. Bundssø–Mjelssø
3. Søborg Sø
5. Sem Sø
6. Spøttrup Sø
8. Lønnerup Fjord
10. Sperring Sø
12. Sundby Sø
14. Tissing Vig
16. Grynderup Sø
18. Lyby Sø (Fårekæret)
20. Iglsø

2. Henninge Nor
4. Bølling Sø
7. Østerild Fj. – Arup Vejle
9. Sjørring Fjord
11. Tastum Sø
13. Sindrup Vejle
15. Legind Vejle
17. Brokholm Sø
19. Torsholm
21. Rødding Sø
23. Bredsgard Sø

25. Fjordmarken
27. Storeø Strand
29. Gråsten Nor
31. Slib Sø
22. Rosborg Sø
24. Gyldenstens Inddæmmede Strand
26. Lumby Inddæmmede Strand
28. Tårup Inddæmmede Strand
30. Hart Sø

Figure 24.2. Investigated drainage projects in Denmark

size of the examined drainage areas was 290 ha, 60 per cent of which was submerged while 40 per cent was low-lying with a high level of ground water before drainage. The largest of the examined reclaimed areas is the Lumby shore on Funen — a drainage scheme comprising 1,000 ha, 580 ha of which is below code zero. The smallest of the reclaimed areas is Lake Sem only covering an area of 12 ha, 7ha of which was formerly a lake.

Agricultural interests

To learn about the agricultural interests, information was gathered on soil conditions, drainage, ownership and configuration of the ground. Table 24.1 shows a survey of factors influencing the agricultural interests. The vast majority of the examined areas are plots owned by many riparian owners each having only a small plot, but there are examples of reclaimed areas where the riparian owners have all their fields and perhaps the buildings situated in the reclaimed area.

The soil is generally of a fine or average quality in the reclaimed areas, whereas the degree of water logging creates some difficulty in some of the reclaimed areas. When considering both type of soil and degree of water logging, it must be concluded that the level of cultivation is good or average in nearly 50 per cent of the examined projects covering about 70 per cent of the area.

Whole or considerable parts of the remaining areas are not well reclaimed. In most cases the drainage condition is sufficient for a rather intensive cultivation of grass and provided the owner can utilize this in his milk production, cultivation of these areas will prove financially sound. If, however, the owners have to use these areas for crop rotation or production of beef cattle, the yield will be modest or none at all. In a few cases the condition of drainage was so poor as to allow only the most extensive utilisation of large parts of the reclaimed area.

The annual drainage costs of the electricity and maintenance averaged 350 kr/ha, which were seldom enough for an area involved in a re-establishment to be considered marginal from an agricultural point of view. In some cases the settling of the drained areas may require new investments in drainage and reduce the profitability of the agricultural use.

Nature interests

Table 24.2 shows a survey of physical factors of importance to the natural conditions after a possible re-establishment. As far as landscape is concerned the areas range from coast stretches, where former inlets have been embanked, to inland areas whose character is entirely different.

When determining the biological interests of a re-establishment the present conditions of nature play an important part. Some of the areas are today of great nature value while others are dominated by agricultural use and are of little if any nature interest. A re-establishment into a lake or an inlet will often result in shallow lakes or inlet areas often surrounded by large low-lying areas.

When re-establishing lakes, water depths of 0.5 to 1.5 m are essential and large parts of the lake must be deeper than 1 m to prevent it from becoming choked. For a number of reasons an adjustment of the level of water will be decisive for the future state of nature. If possible, the level of water should be so as to create a shore full of variations. Islets, protected inlets and small ponds will create the conditions for a rich and varied fauna and flora and will generally contribute to a considerable improvement of the ecological conditions. In most areas the abundance of nutrients is so great that problems of eutrophication are likely to arise unless along with the re-establishment, measures are taken to limit the pollution.

As a whole the nature value of most of the areas will increase to a high degree both from a botanical and zoological point of view. Today certain areas are, however, so valuable that they should be kept fully or in part as they are.

Economic and technical conditions

In most cases purchase of the involved lands will by and large account for the highest expense when re-establishing reclaimed lakes and inlets. In some cases the farmers involved will dispose of such a large piece of land in the reclaimed area as to make the utilisation of the other investments of the farm unprofitable unless compensatory land is found for these farms.

In some cases reclaimed lakes and inlets can be re-established merely by ceasing to pump out the water, but a re-establishment of some of the reclaimed lakes will necessitate a damming of the outlet. Apart from this it may be necessary to carry out other types of work, e.g. on dikes, canals, poles, roads and sewage pipes. The building costs on the five intensively examined areas undergoing a possible re-establishment would account for 10 to 30 per cent of the total costs.

Legislative conditions

Owners of plots will usually be organised in a local reclamation society. The rules of this society will lay down the number of necessary votes for such a society to be abolished. The authorities can, after having bought up the land, open a case for re-establishment, provided that they have got the necessary votes. If there is not a majority for an abolition of the local reclamation society, a re-establishment may be carried through according to the Law of Preservation of Natural Amenities.

For some of the lakes there is no local reclamation society, but re-establishment implies that a damming be carried out on a public

Table **24.1** Agricultural interest in reclaimed lakes and bays

	Area (ha)	Owner*	Soil†	Waterlogging‡	Agricultural¶ value	Annual drainage cost (kr/ha)
FYNS AMTSKOMMUNE						
Gyldenstens indd. Strand	420	1	1,2,2	1	1,2,2	120‖
Einsidelsborg Indd.	826	2	2	1	2,2,2	250
Fjordmarken	920	2–3	2	1	2,2,2	170
Lumby indd. Strand	1,000	1	1,2,3	1	2,2,2	295
Tårup indd. Strand	370	3	2	1	2,2,2	290
Henninge Nor	345	1–2	2	3	2,4,5	
Gråsten Nor	500	3	2,3,3	4	4,4,4	
SONDERJYLLANDS AMTSKOMMUNE						
Slib Sø	350	3	1,4	2	1,1,3	420
Mjellssø	67	3	1,1,3	3	2,3,4	550
Bundssø	157	2–3	1,1,3	2	1,2,3	550
Hartsø	140	2	1,2,4	4	4,4,5	110
ARHUS/RINGKOBING AMTSKOMMUNE						
Bølling Sø	434	2	4	4	4,5,5	0
Sem Sø, lav.	10	2	1,4	3–4	4–5	0
Sem Sø, dyb.	13	2	1,4	3–4	4–5	0
FREDRIKSBORG AMTSKOMMUNE						
Søborg Sø		1–2	1,2	2–3	1,2,3	700
VIBORG AMTSKOMMUNE						
Sjørring Fj.	680	1–2	1,1,2	1–2	1,1,2	
Sperring Sø	165	2	4,4,2	3–4	4,4,3	530
Grynderup Sø	340	3	4,4,2	3–4	4,4,3	150
Brokholm Sø	90	2	4,4,3	3	3,3,4	465
Spøttrup Sø	100	3	4,4,1	3	2,4,5	580
Torsholm	60	3	1,2,4	1	2	100

Fårekaeret	38	2–3	2	3	2,3,4	117
Tastum Sø	800	1–2	1,1,2	2	1,1,2	660
Iglsø	70	3	2	3	2,3,4	82
Rødding Sø	50	1	3,4	3	1,3,4	300
Rosborg Sø	89	3	2,4	4	5,5,5	0
Bredsgard Sø	110		4	4	4,4,5	0
Hovsor Ll Indd.	144	2–3	1,1,3	1–2	1,1,3	520
Hou-Skovsted enge	180	3	1,1,2	1–2	1,1,2	320
Sundby Sø	34	3	4,4,1	3	4,4,3	310
Sindrup Vejle	140	1	1,1,1	1	1,1,1	250
Tissing Vig	82	1	1,1,1	1	1,1,1	
Legind Vejle	59	3	4,4,4	3	3,4,5	520

* (1) Few owners; (2) Medium; (3) Many owners with small plots.
† (1) Loam – organic deposit; (2) Sandy loam – fine sand; (3) Coarse sand; (4) Peat.
‡ (1) No problems; (2) Some problems; (3) Part of the area in grass; (4) The whole area in grass.
¶ (1–3) Good,medium, poor for rotation crops; (4) Allow intensive grass production; (5) Extensive grass production.
‖ Guess.

Table 24.2 The topographic condition after a possible reestablishment

	Water covered			Waterlogged				Salinity condition
	Total area (ha)	Max. water-depth (m)	Over 1.5 waterdepth (%)	Total area (ha)	area<0.7m over water-level (%)	Saltwater possible	Brackishwater possible	Sewage (p.eq)
FYNS AMTSKOMMUNE								
Glydenstens indd.								
Strand	360	0,6	0	.60	50	+	+	70
Einsidelsborg Indd.	160	0,8	0	660	80	–	+	650
Fjordmarken	680	2,5	10–20	240	70	+	+	200
Lumby indd. Strand	580	1,0	0	420	70	+	+	13,000
Tårup indd. Strand	200	0,5	0	170	60	(+)	+	300
Henninge Nor	220	2,3	30	125	70	+	+	50*
Gråsten Nor	380	1,4	0	120	50	+	+	920
SONDERJYLLANDS AMTSKOMMUNE								
Slib Sø	100	1,0	9	250	70	–	+	1,400
Mjelssø	43	3,7	90	24	50	–	+	400
Bundssø	138	4	80	19	50	–	–	200
Hart Sø	50	1,1	0	90	80	–	+	200
ARHUS/RINGKOBING AMTSKOMMUNE								
Bølling Sø	300	1,6	10	134	50	–	–	0
Sem Sø, lav.	6	0,7	0	4	50	–	–	0
Sem Sø, dyb.	8	1,7	33	5	70	–	–	
FREDRIKSBORG AMTSKOMMUNE								
Søborg Sø	430	250	33	180	50	–	–	3,800

VIBORG AMTSKOMMUNE								
Sjørring Sø	460	2	50	220	50	—	—	930
Sperring Sø	120	3	50	45	70	—	—	0
Grynderup Sø	110	1,1	0	230	7,5	—	—	3,300
Bokholm Sø	75	1,8	33	15	50	—	—	0
Spottrup Sø	50	1,6	33	50	50	—	(+)	280
Torsholm	36	1,1	0	24	50	—	—	0
Fårekaeret	20	1,0	0	18	50	—	—	0
Tastum Sø	710	2,1	50	90	50	—	—	170
Iglsø	30	2,4	33	40	33	—	—	0
Rødding Sø	32	2,6	40	18	50	—	—	400
Rosborg Sø	70	2,2	70	19	50	—	—	0
Bredsgård Sø	70	1,5	20	40	60	—	—	?
Hovsor Ll Indd.	52	0,3	0	92	75	+	+	0
Hou-Skovsted enge	60	0,5	0	120	75	+	+	0
Sundby Sø	13	0,8	0	21	50	—	+	600
Sindrup Vejle	100	1,8	40	40	70	+	+	0
Tissing Vig	77	2,6	65	5	100	+	+	390
Legind Vejle	4	0,8	0	55	70	—	+	0?

*Guess

stream. The Law on Water Rights does not allow such a damming unless the owners involved agree. The Law on Preservation of Natural Amenities is, however, considered valid in this case.

Concluding remarks

A re-establishment of reclaimed lakes and inlets will often result in very varied biotopes, and the need for conservation will tend to be less than experienced in other valuable nature areas. A re-establishment of the reclaimed lakes and inlets will therefore generally be attractive from a nature point of view.

It must be stressed, however, that the agricultural interests attached to these areas vary a great deal. In some areas of reclamation the soil is very rich and the agricultural interests may be so important that they are of regional importance, e.g. the southern part of Lolland comprising 15,000 ha of reclaimed area. But there are also areas of reclamation which soon after the re-establishment were given up on account of poor profitability, e.g. Arup–Veslos Vejle in Thy and Sondervig on Mors.

Based on the project 'Re-establishment of reclaimed lakes and inlets in Denmark's. *Skov- og Naturstyrelsen* (part of the Ministry of Environment) (Waagepetersen *et al.*, 1987) suggests re-establishing lakes and inlets on an area covering 20,000 ha of reclaimed areas. The costs of implementing such a scheme will depend on the economy of the production of grass occurring in many of the reclaimed lakes and inlets which are not so well-drained, as well on the general development of the profitability of agriculture

Reference

Waagepetersen J., Mortensen, E., Hansen, J.M., Gyalokay, T., Skotte Moller, H. 1987. Retablering af torlagte soer og fjorde i Danmark. Samlerapport nr. VII vedr. Marginaljorde og Miljointeresser. *Skov- og Naturstyrelsen*, miljoministeriet.

25 Wetland management, agricultural management and nature conservation

C. Newbold, Nature Conservancy Council, Peterborough

This chapter first considers those aspects of agricultural management which affect the survival of flora and fauna within existing undrained primary wetlands. Most primary wetlands in Britain will be sites of special scientific interest (SSSI) (see Appendix 25.A). Second, it considers those management techniques which can sustain wetland interests within a *secondary* wetland area. This is defined as an area which has previously undergone some form of historical drainage but whose pattern of drainage and management are mostly still in sympathy with wildlife, sufficient to make many areas sites of special scientific interest (SSSI).

Definitions are given below of what is meant by wetland and land drainage:

Lowland wetlands: Wetlands do not describe a unified biology but rather a series of discrete ecosystems such as fen, bog, salt marsh, river and lake, which are united by a common physical factor — water. Fundamental to the survival of any primary wetland ecoystem is (i) the amount of water — *water quantity*; (ii) the chemical status of the water — *water quality*; and (iii) the *correct management*. These factors are also important in the survival of relict wetland flora and fauna within a secondary wetland area such as grazing marsh.

Land drainage: A definition of land drainage must include the two component parts; river or arterial drainage and field drainage. It is often difficult to lower the water-table in a field without first either lowering the river level or containing the flow of the arterial channel within flood banks.

Drainage and its potential effect on a primary wetland

Drainage can have five types of impact on wetland wildlife. The first two affect the quantity of water, the remaining three the quality of water.

Water quantity

(i) The most common effect is where water-tables fall, drying out a reserve at its margin. This has happened on several County Trust Reserves (George, 1976). The impact depends to some extent on the permeability of the soils. Peats generally have a low hydraulic conductivity but a drop of 10 cms can have a marked effect on its flora (Gore, 1983).

(ii) More unusually perched water-tables have been breached draining areas outside the scheme (Gosling and Baker, 1980).

Water quality

(iii) Deep drainage has breached saline water-tables (Driscoll, 1983) affecting both freshwater fauna and flora and yielding unsuitable water for irrigation.

(iv) The deepening of drains has exposed pyritic peats causing ferrous rich 'tides' of water (Bloomfield, 1972). These affect invertebrates, smother aquatic plant-growth, and reduce the penetration of light (NCC File, 1982). They also cause pollution and increasing sedimentation in the main river.

(v) Sulphide-rich peats have been exposed which through oxidation/reduction processes release sulphuric acid. Oscillations in pH between 7.4 and 3.2 have had disastrous effects on the flora and fauna of Calthorpe Broad, Norfolk (Gosling, 1971; Gosling and Baker, 1980). Small areas of West Sedgemoor, Somerset, have suffered similar problems (Williams, R.J., personal communication, 1982).

Ameliorative and remedial management

Certain ameliorative measures can be undertaken; but sometimes the cost prohibits action. Perched water-tables can be protected by 'clay walling', or sealing in the water-table. This has been achieved on a 200 ha reserve at Woodwalton Fen NNR, Cambridgeshire, at an approximate cost of £0.4 m. This was paid for by the Middle Level Commissioners (a local drainage board) with a grant from the European Economic Community (EEC). Pumps feeding iron-rich water into Martham Broad, part of the Hickling Broad NNR, Norfolk, have been moved so that the water now bypasses the reserve. This cost to the Nature Conservancy Council was £7,000 in 1983–4 (NCC Tenth Annual Report, 1984). Presumed intrusion of saline water into Hickling Broad NNR has had to be accepted. Water, rich in sulphuric acid, exposed through deep drainage on a small area of West Sedgemoor SSSI, prior to the establishment of the SSSI, has to be accepted. Fluctuations in acidity at Calthorpe Broad NNR, Norfolk, are crudely treated by liming.

One way of avoiding such effects would be to survey the area prior to any drainage work for its soil types, hydrology and wildlife interest, and to produce drainage schemes which are sensitive to these environmental conditions (Newbold, 1986). By optimising water-tables irrigation requirements would be reduced. Some Internal Drainage Boards (IDBs) are attempting to do just this on existing schemes.

River Engineering

One of the most damaging forms of drainage activity was that of river engineering to serve the needs of land drainage. The river engineer's aim was and is generally to minimise both the quantity of water which flows onto the land, and the extent and period of inundation. Channel capacity is increased and obstacles are removed which contribute to the roughness coefficient of the river. Projecting bushes and trees, islands, sand bars and lush aquatic growth, all have been removed in an effort to create the most economical shape to maximise water flow — the trapezoidal channel.

The destruction of habitats in rivers to serve the needs of land drainage increased during the 1960s and 1970s. The lowland rivers of England and Wales suffered most. The Great Ouse, the rivers Welland and Nene flow through flat, fertiles, river valleys before they help drain the fens of Cambridgeshire; 44 per cent and 47 per cent of their respective lengths have been channelised (Brookes, 1981, and personal communication 1984). In contrast, only 11 per cent of all rivers in the mountainous county of Cumberland (Cumbria) have been channelised.

There are no figures available for Scotland, but channelisation is organised, generally, on a need basis by the setting-up of a consortium of farmers wanting to 'improve' their local stretch. The water authorities of Scotland are distinct in function and law (Water (Scotland) Act 1967) from the English and Welsh water authorities. The water authorities in England and Wales exercise a general supervision over all matters relating to land drainage. They are responsible for engineering works on 'main' rivers. In general, it is safe to state that the rivers in Scotland have not been modified to the degree they have in England and Wales.

Ameliorative and remedial management
The conservationist, with his knowledge of the habitat requirements of plants and animals, has in the last ten years been assisting the water authorities of England, Wales and Scotland in more effectively furthering the nature-conservation interest in rivers. Equally the river engineers

have shown an enthusiasm and ingenuity in the way they have built in conservation considerations to their land-drainage brief.

There are numerous guidelines bringing together these once mutually exclusive activities. The most significant are those published by the Water Space Amenity Commission (1981), the Nature Conservancy Council (Newbold *et al.*, 1983) and the Royal Society for the Protection of Birds in conjunction with the Royal Society for Nature Conservation (Lewis and Williams, 1984).

In many water-authority areas such sympathetic management of rivers is very much the issue of, 'too little, too late'. However, some water authorities are beginning to recreate river habitats where they once destroyed the 'naturalness' of the river to create their trapezoidal channel.

Drainage and its potential effect on seconary wetlands

The grazing marsh or secondary wetland is the most common form of wetland in Britain. It describes a former wetland where drainage ditches and their margins contain a relict wetland flora and fauna. These ditches often delineate the field boundary and provide a wet fence to contain cattle within an often poorly drained field. Historically these areas may have been the most difficult areas to drain for arable cropping and this remains true today for those few still remaining.

Many grazing marshes have contracted in size over the last three or four decades due to improvement in both field and arterial drainage (Mountford and Sheail, 1982, 1984 and 1985). Fragments of true grazing marsh survive in, for example, the Pevensey Levels, Sussex; the Romney Marsh, Kent; places such as the Gwent Levels, South Wales; the Somerset Levels and Moors; and the Broadland of Norfolk.

Ameliorative and remedial management
Within all these areas management has been more or less in sympathy with wildlife. The poorly drained fields are bordered by ditches with a high and relatively stable water level and the amount of water does rebuff to some extent the pollution which can result from fertiliser run-off. Nowadays aquatic plants are very rarely hand cut but mechanical cutting and certain herbicides can, when used correctly, still produce a diverse flora and fauna. Appropriate management methods have been detailed elsewhere (Newbold, 1974, Newbold, 1976, Newbold 1984). Ideally only one of the four ditches surrounding a field should be 'cleansed' annually. Ditches are thus cleaned on a four-year rotation. If this is not practical then wetland bankside and aquatic strips could be left on one side of the ditch and these strips managed on a four-year rotation.

There is every advantage for colonising invertebrate fauna such as beetles and dragonflies if the southern, south-eastern and south-western drain edge is kept free of vegetation. Insects generally colonise from the southern edge. If the drain is facing north to south then either side can be cleansed.

Irrigation

Land drainage, the movement towards arable crops, and the hot dry summers of 1976 and 1984 have increased demands for irrigation water. The number of abstraction licences consented by Anglian Water for irrigation purposes rose from 2,897 in 1969, to 2,950 and 3,382 in 1976 and 1983, with an increase in abstraction volume from 155.9 to 172.9 to 224.3 X $10^3 m^3/d$ respectively (Anglian Water Authority, 1983). In the grazing marsh districts of Romney (Kent), Broadland (Norfolk), and the Gwent Levels (South Wales), changes to arable have resulted in the use of ditch water for irrigation. These ditches once provided wet fences for cattle by delineating field boundaries. The stable water levels provided a rich habitat for aquatic wildlife. The fluctuating spring water levels found in ditches draining arable crops and intensive grass leys, and the drying out of such ditches in summer have destroyed aquatic wildlife (Mountford and Sheail, 1982).

Within the Anglian Water Authority area, irrigation demands have been so high from bore holes sunk in the chalk aquifer that fen water-tables have fallen several kilometres distant from the abstraction point (Lloyd, J., personal communication, 1987).

Eutrophication and its effects on primary wetland and secondary wetland wildlife

The most pervasive chemical effect on a wetland is that of eutrophication. Approximately 50 per cent of the enrichment arises from sewage sources under the control of water authorities, approximately 30 per cent arises from industry and approximately 20 per cent arises from agriculturally improved land. All sources can pollute water entering a reserve or grazing marsh. The most common agricultural pollutants are fertiliser run-off, silage liquor and animal slurry. These stem from recent changes in agricultural practice, particularly the use of animal slurries to treat the land with fertiliser and pollution problems arising from increased silage storage. This has meant that once relatively unpolluted salmonid rivers and lakes are now seriously polluted. Agricultural eutrophication although only some 20 per cent of the total is now polluting some of the once pristine rivers and lakes.

Eutrophication describes the effect of increased plant nutrients on wetland plant and animal life. The first effect is one of increased plant production and is very quickly followed by smothering growths of algae, and loss of salmonid fish stocks. Their full effects are described by Lund (1972) and Moss (1983). The effects of eutrophication are difficult to reverse and it is best to try and prevent the effect occurring through appropriate management.

Appropriate wetland management and research necessary to effect better management

Land drainage

(i) The results of corridor surveys of riverine wildlife are now incorporated into the design and maintenance phase of most river engineering schemes such that the needs of wildlife are considered alongside the drainage needs. There is a need to identify those bankside habitats which are most beneficial to wildlife so that the correct ones are conserved.

(ii) There is a need to validate the advantage which could accrue for wildlife, for the drainage engineer and farmer by optimising and stabilising water-tables in drainage channels during the summer.

Eutrophication (enrichment) and pesticides usage

(i) NCC is conducting research on the effectiveness of grass strips alongside field drains, in an attempt to counter some of the fertiliser run-off problems. We are also researching the effectiveness of 'lenses' of reed around tile drain outlets. Only two sites are being studied. The .programme could be expanded.

(ii) There is a need to research an effective and cheap method of removing the potential pollutants from animal slurry and silage liquor. The root zone method (Boon, 1985) could be used but its efficiency needs to be researched.

(iii) There is a need to develop an accurate fertiliser spreader and/or modify existing machinery.

(iv) There is a need to develop more accurate pesticide spraying machinery and/or modify existing machinery to reduce the impact of pesticide drift.

(v) The effects of pesticide drift on wetland flora and fauna needs to be researched.

References

Bloomfield, C. 1972. The oxidation of iron sulphides in soils in relation to the formation of acid sulphate soils and of ochre deposits in field drains. *Journal of Soil Science, 23,* 1–16.

Boon, A.G. 1985 Report of a visit by members and staff of WRC to Germany GFR to investigate the root zone method for the treatment of waste waters Water Research Centre, Stevernage, Hertfordshire pp 1–65.

Brookes, A. 1981, 'Channelization in England and Wales'. Discussion Paper No. 11. Dept of Geography, The University of Southampton.

Driscoll, R.J. 1983. Broadland Dykes: The loss of an important Wildlife Habitat. *Transactions of the Norfolk and Norwich Naturalist Society, 26*, 170–2.

George, M. 1976. Is wildlife being drained away? *Conservation Review, 12*, Spring. published by Society for the Promotion of Nature Reserves, Lincoln.

Gore, A.J.P. ed. 1983. *Ecosystems of the world, Mires: Swamp, Bog, Fen and Moor.* ITE Monks Wood Experimental Station, Abbots Ripton, Hungtingdon, Cambs.

Gosling, L.M. 1971. 'Calthorpe Water acidity — background, effects, cause and suggested treatment'. Unpublished report. Nature Conservancy Council, Norwich.

Gosling, L.M. and Baker, S.J. 1980. Acidity Fluctuations at a Broadland site in Norfolk. *Journal of Applied Ecology, 17*, 479–90.

Lewis, G. and Williams, G. 1984. *Rivers and Wildlife Handbook:A Guide to Practices Which Further the Conservation of Wildlife on Rivers.* Royal Society Protection of Birds, Royal Society Nature Conservation Sandy, Bedfordshire, pp 295.

Lund, J.W.G. 1972. Eutrophication. *Proceedings Royal Society Land, 180*, 371–82.

Moss, B. 1983. The Norfolk Broadland: Experiments. *Biol. Rev., 58* 521–61.

Mountford, J.O. and Sheail, J. 1982. 'The impact of land drainage on wildlife in the Romney Marsh. The availability of baseline data'. Institute of Terrestrial Ecology. Project No. 781. Interim report to the Nature Conservancy Council.

Mountford, J.O. and Sheail, J. 1984. 'Plant life and the watercourses of the Somerset Levels and Moors'. Institute of Terrestrial Ecology Project No 718. Interim report to the Nature Conservancy Council.

Mountford, J.O. and Sheail, J. 1985. 'Vegetation and changes in farming practice on the Idle/Misson Levels, (Nottinghamshire, Humberside and South Yorkshire)'. Institute of Terrestrial Ecology Project No. 718. Interim report to the Nature Conservancy Council.

Nature Conservancy Council File 1982. 'Internal Report: The effects of iron ochre on freshwater plants and invertebrates within grazing marsh systems in Norfolk'. East Anglia Regional Office, 60, Bracondale, Norwich.

Nature Conservancy Council 1984. *Tenth Annual Report*. HMSO, London.

Newbold, C. 1974. The ecological effects of the herbicide dichlobenil within pond ecosystems. Proceedings of the European Weed Research Council Fourth International Simposium on Aquatic Weeds, Vienna. pp. 37–52.

Newbold, C. 1975. Herbicides in Aquatic Systems. *Biological Conservation*, 7, 97–118.

Newbold, C. 1976. *The Influence of Aquatic Herbicides on Wetlands* Council of Europe European Information Centre for Nature Conservation Wetlands Campaign.

Newbold, C. 1984. Aquatic and bankside herbicides. In *Rivers and Wildlife Handbook* RSPB, RSNC. Joint Publication, Royal Society for the Protection of Birds, Sandy, Bedfordshire, pp. 209–14.

Newbold, C. 1986. 'Nature conservation, river engineering and the drainage area of a river scheme'. *International Commission for Irrigation and Drainage (ICID) Bulletin*, 35, 30–6.

Newbold, C., Pursleglove, J. and Holmes, N.T.H. 1983. *Nature Conservation and River Engineering*. Nature Conservancy Council, Peterborough.

Water (Scotland) Act 1967. HMSO Edinburgh.

Water Act 1973. HMSO London.

Water Space Amenity Commission 1980. *Conservation and Land Drainage Guidelines. Drainage Flood Protection and Sea Defence Work, England and Wales*. WSAC, London.

Wildlife and Countryside Act 1981. HMSO London.

Appendix 25.A: Sites of special scientific interest and the Nature Conservancy Council's role in the remainder of the countryside

The Nature Conservancy Council (NCC) is the statutory body which promotes nature conservation in Great Britain. It gives advice on nature conservation to government and all those whose activities affect the wildlife and wild places of Great Britain. It also selects and manages a series of National Nature Reserves (NNRs). The NCC has also identified other important places known as Sites of Special Scientific Interest (SSSI). In these areas the Wildlife and Countryside Act, 1981 (HMSO, 1981) requires owners and occupiers to tell the NCC in advance if they intend to carry out a damaging operation, so that the NCC can discuss an alternative, less-harmful approach, or offer financial help within a nature conservation management agreement.

Although it is very important to conserve special sites it must also be ensured that the remainder of the countryside is managed sympathetically for wildlife. Few species can survive only in isolated sites, most rely on the countryside around them. The NCC's role in the wider countryside is one of advice and persuasion.

The Wildlife and Countryside Act, 1981, also specifically protects a number of wild plants and animals, so the NCC, in the wider countryside, attempts to prevent the killing, injury, taking, or selling of specially protected animals.

Conclusions and Recommendations

Conclusions and Recommendations

It was opportune that the Programme Committee for Land and Water Use and Management arranged for the topic of Agricultural Management and Environmental Objectives to be considered by a Workshop at a time when environmental issues are to the fore in Community policy and public perception. *Alternative land use* is an important aspect of Community agricultural policy; environmental objectives provide potential opportunities for alternative land use but environmental concerns also need to be a component in policy development. Thus there is much concern that development of forestry as an increasing land use should take due account of environmental aspects. The watchword is integration; it is essential to take account of the full range of interests and agree objectives at the stage of conception. This is equally relevant for developments outside agriculture, such as recreation and industry, but which may affect both agriculture and the environment.

National Reports, technical papers and the discussion at the Workshop highlighted the variety of relevant measures available in Member States, ranging from the legal protection of nature reserves to voluntary measures to safeguard areas of scientific and landscape value. Much interest was focused on the recently established Environmentally Sensitive Areas, not least due to the visit to a recently designated ESA. The range of national measures available clearly affects the extent and approach to protection and hence the identification of research topics. This Workshop was an effective way to share experience between Member States, and the Commission should take further opportunities to use this approach in a selective manner.

The Workshop reviewed relevant research already under-way in Member States, some of which was previously unknown to participants. It is necessary to ensure that there is a greater mutual awareness of relevant work in Member States. There is also a need to draw together and disseminate information. Without this there is no cross-fertilisation of ideas; equally important, scientists are less able to reinforce their conclusions by reference to similar work and avoid the mistakes of others in a rapidly developing and potentially controversial subject area.

Gaps in the research effort have been highlighted in the introductions to the various parts of the book; these include specific aspects of research which could be covered by extensions of Community or Member States programmes. However, the Workshop did emphasise the importance of closer collaboration between research teams and there is considerable scope for the Community to encourage collaboration between research teams working in different geographical locations in order to develop a truly Community information base. The Workshop recommends that:

The Commission should:

(1) take steps to ensure that relevant research information on environmental topics in agriculture in Member States is drawn together on a Community-wide basis, and procedures developed for its dissemination;
(2) encourage through national agencies the development of more effective ways of advising farmers of the benefits of ecologically sympathetic management systems;
(3) encourage closer collaboration between research teams either working to the same objectives in different geographical areas, e.g. grassland management to increase species diversity, or taking different approaches to the same research problem, e.g. biological and landscape studies of field margins;
(4) in the context of research on pollution of drinking water by nitrate and pesticides, ensure that the wider aspects of pollution of water, particularly eutrophication, and its effect on other environmental interests are included.

Successor workshops should cover:

(1) management of grassland for environmental and agricultural objectives and for situations where environmental objectives are dominant, in order to review the state of knowledge and identify research needs on a Community basis;
(2) the conservation and agricultural benefits of pest control by encouraging populations of beneficial insects in non-crop land, including field margins;
(3) problems facing Mediterranean countries, since the topics considered in the present Workshop were mainly relevant to Northern European countries.

Research is required on:
(1) the opportunities to conserve wildlife where grassland is intensively managed for agricultural production;
(2) the autecology of species of plants and animals in managed grassland and wetland, and of endangered communities of plants;
(3) the economic cost of management systems to meet specific environmental objectives;
(4) the extension of current studies of field margins to encompass crops other than arable, particularly grassland.

The Workshop was a valuable initiative; it has acted as a catalyst to collaboration between participants but this needs to be developed into closer collaboration on a Community basis. Opportunities and avenues have been identified; only the Community and Member States can take this forward.

Annexes

Appendix I: Agricultural Management and Environmental Objectives SCAR Workshop, 14–17 July 1987 — Programme, 14 July

Welcome

Introductory Talk — B. *Halliday* (Farmer, Exmoor)

15 July

Chairman — P. *Needham* (MAFF)

Review Paper — The Efficiency of the Agricultural Industry in relation to the Environment

P.B. *Tinker* (Natural Environment Research Council, UK)

Discussion

Session 1 Management of Field Margins

Hedgerows and hedgerow networks as wildlife habitat in agricultural landscapes

J. *Baudry* (France)

The Status of hedgerow field margins in Ireland

R. *Webb* (Ireland)

Agriculture changes in scrub and grassland habitats in Europe

P. *Devillers* (Belgium)

The dispersal of plants from field margins

J. *Marshall* (Agriculture and Food Research Council, UK)

The role of field margins as reservoirs of beneficial insects

S. *Wratten* (University of Southampton, UK)

Chairman — P.B. *Tinker* (Natural Environment Research Council, UK)

Review paper — Monitoring of Landscape and Wildlife Habitats

A.J. *Hooper* (MAFF, UK)

Session 2 — Management of Grassland

Utilisation of peatland for grassland and wetland habitat — contrast or agreement

H. *Kuntze and J. Schwaar* (West Germany): delivered by J. Schwaar

Bird protection in grassland in the Netherlands

L.A.F. *Reyrink* (The Netherlands)

Investigation on the arthropod fauna of grasslands

J. *Maelfait* (Belgium)

Distribution and management of grassland in the United Kingdom

A. *Hopkins* (Agriculture and Food Research Council, UK)

The effects of agricultural change on the wildlife interest of lowland grasslands

T.C.E. *Wells and J. Sheail* (Natural Environment Research Council, UK): delivered by J. Sheail

Demonstration of Ministry of Agriculture videos, display material etc

16 July Visit to Somerset Levels and Moors

West Huntspill: management of an arterial drainage system, involving pumped drainage, and its relationship to a tidal river

D. *Alsop*, Principal Engineer, Wessex Water Authority

Burrow Mump: traditional farming on the enclosed moor and intensified farming adjacent to it

J. *Row and J. Shove*, ADAS, Taunton

West Sedgemoor: designation as an SSSI, continuation of farming, part played by Royal Society for Protection of Birds and local liaison
Dr B. Johnson, Nature Conservancy Council
Tadham Moor: description and discussion of field project
F. Kirkham, Animal and Grassland Research Institute, Agriculture and Food Research Council
Ecology and Agriculture of Tadham Moor
Dr J.M. Way and J. Shove, ADAS

17 July

Introduction
Chairman — *E. Culleton* (EC Commission)
Session 3 — Management of Wetlands
Applied ecological research on the conservation of wet grasslands in relation to agricultural land use in Flanders, Belgium
E. Kuijken (Belgium)
Plant production and agricultural use of wet lowland and dry sandy soils in Denmark
L. Hansen (Denmark)
Topsoil compaction causes ponding in grazed pastures
J. Mulqueen (Ireland)
Reclaimed Danish lakes and inlets — agricultural value and nature interests
J. Waagepetersen (Denmark)
Wetland management, agricultural management and nature conservation
C. Newbold (Nature Conservancy Council, UK)

Discussion

Round table discussion
Chairman — *J.R. Park* (MAFF)
    Rapporteur report (Session 1) — *C. Feare* (MAFF)
    Rapporteur report (Session 2) — *J.M. Way* (MAFF)
    Rapporteur report (Session 3) — *I.M. Tring* (MAFF)

General Discussion

Conclusion

Belgium
Dr P. Devillers
Chef de travaux
Institut Royal des Sciences Naturelles de Belgique
28, rue Vautier
B 1040 Brussels

Dr E. Kuijken
Director of the Instituut voor Natuurbehoud
Kiewitdreef 3
B 3500 Hasselt

Dr J.P. Maelfait
Ministerie van de Vlaamse Gemeenschap
Administratie voor Ruimtelijke Ordening en Leefmilieu
Instituut voor Natuurbehoud
Kiewitdreef 3
B 3500 Hasselt

Denmark
Forstander, Lic. agro. Lorens Hansen
Landbrugscentret
Statens Forsogsstation
Flensborgvej 22
Jyndevad
DK — 6360 Tinglev

Jesper Waagepetersen
Hedeselskabet
Danish Land Development Service
PO Box 110
8800 Viborg

France
Dr J. Baudry
Institut National de la Recherche Agronomique
Systèmes Agraires et Développement
Le Robillard, 14170 St-Pierre-sur-Dives

Ph. Merot
Laboratoire de Recherches de Science du Sol
INRA-ENSA
65 Rue de Saint Brieuc
F-35042 Rennes Cedex

Germany
Dr A. Krause
Bundesforschungsanstalt für Naturschutz und Landschaftsokologie
Konstantinstr 110
B 5300 Bonn 2
(Telex 885 790)
Tel. 0228/84910

Dr J. Schwaar
Niedersächsisches Landesamt für Bodenforschung-Boden-
technologisches Institut
Friedrich-Missler Strasse 46/50
D 2800 Bremen 1
Tel. 0421/230089 (no telex)

Ireland, Republic of
J. Mulqueen
Agricultural Institute
Ballinrobe
C. Mayo

R. Webb
An Foras Forbartha
St Martin's House
Waterloo Road
Dublin 4

Luxembourg
Mr J. Frisch
Administration des Services Techniques de L'Agriculture
Boite Postale 1904
16, Route de'Esch
1019 Luxembourg

Netherlands
Dr M. Hoogerkamp
Centrum voor Agrobiologisch Onderzoek (CABO)
Bornsesteeg 65
PO Box 14
6700 AA Wageningen

Dr L.A.F. Reyrink
Ministry of Agriculture and Fisheries
Bureau for Land Management
Griffioenlaan 2
PO Box 20022
3502 LA Utrecht

Portugal
Engo Eugenio Sequeira
Est Agron Nacional
2780 Oeiras

Engo Augusto Rodriques
Est Agron Nacional
2780 Oeiras

EC Commission
Dr E. Culleton
Commission of the European Communities
Rue de la Loi 200
1049 Brussels

United Kingdom
Dr J. Park
Ministry of Agriculture, Fisheries and Food
Great Westminster House
Horseferry Road
London
SW1P 2AE

Mr P. Needham
Ministry of Agriculture, Fisheries and Food
Great Westminster House
Horseferry Road
London
SW1P 2AE

Mr B. Halliday
Farmer
Ashton
Countisbury
Lynton
N. Devon
EX35 6NQ

Dr P.B. Tinker
Director of Science,
Terrestrial and Freshwater Sciences
Natural Environment Research Council
Polaris House
North Star Avenue
Swindon
SN1 1EU

Dr E.J.P. Marshall
Institute of Arable Crops Research
Department of Agricultural Sciences
University of Bristol
Long Ashton Research Station
Long Ashton
Bristol
BS18 9AF

Dr S.D. Wratten
Department of Biology, Southampton University
Building 44
The University
Southampton
SO9 5NH

Mr A.J. Hooper
Ministry of Agriculture, Fisheries and Food
Great Westminster House
Horseferry Road
London
SW1P 2AE

Mr A. Hopkins
Institute for Grassland and Animal Production
Grassland Utilisation Department
North Wyke Research Station
Okehampton
Devon
EX20 2SB

Dr J. Sheail
Institute of Terrestrial Ecology
Natural Environment Research Council
Monks Wood Experimental Station
Huntingdon
Cambridgeshire
PE17 2LS

Dr C. Newbold
Nature Conservancy Council
Northminster House
Peterborough
PE1 1UA

Dr C.J. Feare
Ministry of Agriculture, Fisheries and Food
Worplesdon Laboratory
Tangley Place
Worplesdon
Surrey
GU3 3LQ

Dr J.M. Way
Ministry of Agriculture, Fisheries and Food
Great Westminster House
Horseferry Road
London
SW1P 2AE

Mr I.M. Tring
Ministry of Agriculture, Fisheries and Food
Great Westminster House
Horseferry Road
London
SW1P 2AE

Dr J.C. Sherlock
Ministry of Agriculture, Fisheries and Food
Great Westminister House
Horseferry Road
London
SW1P 2AE

Mr M. Taylor
Countryside Commission
John Dower House
Crescent Place
Cheltenham
Gloucestershire
GL50 3RA

Department of the Environment
Tollgate House
Houlton Street
Bristol
BS2 9DJ
(represented by Miss R. Ashman, Mr D. Brown, Mr M. Lloyd and
Mrs S. Toland).